D0952128

Out There

OUT THERE

*A Scientific Guide to Alien Life,
Antimatter, and Human Space
Travel (For the Cosmically
Curious)*

Michael Wall, PhD

Senior Writer, Space.com

GRAND CENTRAL
PUBLISHING

NEW YORK   BOSTON

Grand Central Publishing
Hachette Book Group
1290 Avenue of the Americas, New York, NY 10104
grandcentralpublishing.com
twitter.com/grandcentralpub

First Hardcover Edition: November 2018

Grand Central Publishing is a division of Hachette Book Group, Inc. The Grand Central Publishing name and logo is a trademark of Hachette Book Group, Inc.

The publisher is not responsible for websites (or their content) that are not owned by the publisher.

The Hachette Speakers Bureau provides a wide range of authors for speaking events. To find out more, go to www.hachettespeakersbureau.com or call (866) 376-6591.

Library of Congress Cataloging-in-Publication Data

Names: Wall, Michael (Biologist), author.
Title: Out there : a scientific guide to alien life, antimatter, and human space travel (for the cosmically curious) / Michael Wall, PhD (senior writer, Space.com).
Description: New York : Grand Central Publishing, [2018] | Includes index.
Identifiers: LCCN 2018022905| ISBN 9781538729373 (hardcover) | ISBN 9781549168482 (audio download) | ISBN 9781538729380 (ebook)
Subjects: LCSH: Life on other planets. | Extraterrestrial beings. | Outer space–Exploration.
Classification: LCC QB54 .W178 2018 | DDC 576.8/39–dc23
LC record available at https://lccn.loc.gov/2018022905

ISBNs: 978-1-5387-2937-3 (hardcover), 978-1-5387-2938-0 (ebook)

Printed in the United States of America

LSC-C

10 9 8 7 6 5 4 3 2 1

For Teddy

Table of Contents

Introduction   1

Part I: What's Out There?

1. Where Is Everybody?   9

2. Are We All Martians?   27

3. What Does ET Look Like?   40

4. Do Aliens Have Sex?   66

5. What Are We Looking For?   74

6. Where Is ET Hiding?   98

7. How Would the World Be Told?   124

8. Could We Talk to ET?   133

9. How Would the World Respond?   142

10. Have We Already Found ET?   153

11. Will Aliens Kill Us All?   168

Part II: Getting Out There

12. Will We Colonize the Moon and Mars?   183

13. Can We Go Interstellar?   200

14. Will There Be a *Homo spaciens*?   215

15. Is Time Travel Possible?   222

16. What Will Happen to Us?   229

Bibliography   235

Index   237

Acknowledgments   244

About the Author   246

Out There

Wow!

Two days after Elvis Presley died, astronomer Jerry Ehman was sitting at his kitchen table, poring over an eye-straining thicket of numbers and letters on computer printouts. This was not his way of coping with the King's untimely demise; it was his job. Well, sort of. Ehman had volunteered to look for interesting patterns in this messy mass of data, which had been collected by Ohio State University's Big Ear radio telescope.

Something in the jumble caught Ehman's eye: the vertical string *6EQUJ5*. It was so surprising that he circled the mini column with his red pen and wrote "Wow!" in the margin in a lovely, looping script. It's nice that this iconic moment was immortalized in such an eye-pleasing way. The story would be marginally worse if the word had been written in my mom's pinched and slanting scrawl.

6EQUJ5 was code describing a radio signal that had come in three days previously, on August 15, 1977. Ehman saw that the Wow! signal, as it has come to be known, was strong, covered a narrow range of wavelengths, and lasted 72 seconds, as would be expected from a deep-space source. (This was the length of time that the Big Ear could observe a distant cosmic target, before Earth's rotation rolled another patch of sky into view.) All of these characteristics were consistent with a transmission from an alien civilization.

But there was more. The Wow! signal's frequency was 1,420 megahertz—squarely within the "water hole," the slender, radio-quiet range that many astronomers had predicted ET would use to contact us. The name stems from its position between the frequencies of naturally occurring cosmic hydrogen atoms (H) and hydroxyl (OH) molecules, which together form water. Also, it's a kind of joke: the water hole should attract conversation, just like an office water cooler. (If you're a fan of nature shows, *water hole* may evoke images of wildebeest getting ambushed by lions or crocodiles at dwindling, muddy pools, which is a valid image as well, if ET means to do us harm.)

This was very intriguing indeed, but astronomers

would need more information before they could say anything definitive about the Wow! signal. For starters, they'd need to observe it again. So Ehman and his colleagues tried with the Big Ear, again and again. Nothing. Other astronomers sought to pick it up as well, using a variety of different scopes. Silence. Researchers have continued this effort over the decades, and nobody has had any luck. The Wow! signal was a one-off, a single cry in the dark.

So what was it? A weird, isolated natural event? Terrestrial interference somehow masquerading as a deep-space signal? Or, hope against hope, a "hello" beamed across the dark and frigid depths by ET?

"I don't think it can ever be determined," Ehman said. "I'm frustrated that I can't draw any further conclusions than I already have."

Though the Wow! signal remains a cosmic unicorn, a lot has changed since its detection four decades ago. Back then, the only planets astronomers knew about circled the sun. Scientists generally thought that life was as soft as a trust-fund kid, restricted to a narrow range of temperatures, pressures, and pH.

Now we know that the universe harbors more planets than stars, and huge numbers of these alien worlds probably look a lot like ours. But even if we never found an Earth twin, that wouldn't necessarily mean we're alone in the universe. Microbiologists have discovered that many of our planet's smallest inhabitants are unbelievably tough, scratching out a living in boiling-hot mud pots, frigid pools buried beneath Antarctic ice, and soda lakes so alkaline they'd eat the skin off a flamingo's legs. The possible abodes for life are nearly endless.

The search for ET has moved from the sci-fi fringes into the mainstream and onto the front pages of newspapers around the world. Scientists are looking hard now, and their efforts are about to ramp up, as a new generation of powerful telescopes, both on the ground and in space, is about to come online. Excitement and optimism are in the air.

"I think it'd be the biggest surprise if we don't find something," said Lisa Kaltenegger, an associate professor of astronomy at Cornell and director of the university's Carl Sagan Institute. Kaltenegger is focused primarily on finding biosignatures in the atmospheres of alien worlds—generally speaking, evidence of "simple" life such as microbes.

But Ehman is just as bullish on intelligent aliens. "I'm convinced that there are probably millions or billions of extraterrestrial civilizations, if we include all of the galaxies in the universe," he said.

As the search for ET heats up, so too does the push to expand humanity's footprint out into the solar system.

NASA is planning to return astronauts to the moon and then get them to Mars; the European Space Agency is really excited about setting up an international "moon village." But much of the action is in the private sector. SpaceX and Blue Origin—led by billionaires Elon Musk and Jeff Bezos, respectively—are launching, landing, and reflying rockets, showcasing tech that could slash the cost of spaceflight enough to get us off this rock in a meaningful way for the first time ever.

Musk wants to set up a million-person city on Mars. Bezos wants to get millions of people living and working in space, in so-far unspecified locations that presumably do not include the surface of the sun, the interiors of black holes, or the kitchens of Romulan warships.

A number of companies aim to start mining asteroids and the moon in the next few years. And we're already manufacturing stuff off-Earth, using machines launched to the International Space Station.

"A thousand years in the future, looking back, people will see this as the inflection point, where we did transition to a multiworld and spacefaring civilization," said Bob Richards, the CEO of Moon Express, a company that aims to provide robotic transportation services to the lunar surface and mine the moon and a number of other bodies throughout the solar system.

There's a lot going on out there and a lot out there to

see. In this book, we're going to take a little tour of the great beyond, asking some pertinent (and some impertinent) questions about ET: Does he/she actually exist? If so, why so shy? Do aliens have sex? And we'll touch on humanity's push for another giant leap: Will we colonize Mars? Can we go interstellar? Could we really travel back in time and stab Hitler in the neck with an awl?

Enough of the preliminary chatter. Let's get started!

Part I

What's Out There?

Chapter 1

Where Is Everybody?

In 1950, Nobel Prize–winning physicist Enrico Fermi—who led the team that created the first-ever nuclear reactor, the inadequately named Chicago Pile-1—and a few of his colleagues were discussing UFOs during their lunch break. The conversation prompted Fermi to ask his companions, "Where is everybody?"

Fermi meant that the lack of visits by ET is distinctly odd. The Milky Way harbors hundreds of billions of stars and is about 13 billion years old, so there has been plenty of time and opportunity for alien civilizations to rise and spread throughout the galaxy. By some estimates, a colonization-minded species with propulsion technology not much more advanced than our own could island-hop its way to every corner of the Milky Way in just a few million years.

The physicist's simple question is enshrined now as Fermi's paradox—one of the two coolest paradoxes of all time, along with the crocodile paradox—and it continues to puzzle scientists to this day. Indeed, the mystery has deepened considerably over the years. For one thing, we're not just talking about the lack of visitation anymore. In 1960, 6 years after Fermi's death, astronomer Frank Drake pointed a radio telescope at West Virginia's Green Bank Observatory at the nearby sun-like stars Tau Ceti and Epsilon Eridani, kicking off the search for extraterrestrial intelligence (SETI).[1] Nearly 60 years later, SETI scientists are still hunting for the first confirmed peep from ET.

Then there's the exoplanet revolution. Alien worlds were purely hypothetical objects in Fermi's day and for decades afterward; scientists didn't announce the first

---

1   In 1961, Drake came up with a way to estimate the number of alien civilizations that may be active and communicating right now throughout the galaxy. The famous Drake equation takes into account star-formation rates and key variables such as the fraction of stars that host planets; the proportion of those planets that could support life; the number that actually do host life, and how many of those planets end up supporting intelligent, communicating civilizations; and how long those civilizations keep beaming signals out into the cosmos. We don't know many of these numbers, of course, so estimates derived from the Drake equation can vary widely.

confirmed detection of a planet beyond the solar system until 1992. But in the last decade or so, NASA's Kepler space telescope and other instruments have revealed that the cosmos is teeming with possibly life-supporting worlds. Kepler's discoveries suggest that about 20 percent of the Milky Way's sun-like stars host an Earth-sized world in the "habitable zone"—that just-right range of orbital distances that would allow you to walk around in flip-flops pretty much year-round. The proportion appears to be similar for red dwarfs, the small, dim stars that dominate our galaxy. (About 75 percent of Milky Way stars are red dwarfs, whereas just 10 percent or so are similar to our sun.)

"There's a lot of real estate out there, and we now know that," said radio astronomer Jill Tarter, who cofounded the SETI Institute in Mountain View, California, and served as the inspiration for Ellie Arroway, the lead character in Carl Sagan's novel *Contact* and the movie based on it.

Not all of this real estate is way out in the boonies, either. The sun's nearest neighbor, the red dwarf Proxima Centauri, hosts an Earth-size planet in the habitable zone. Seven rocky planets circle the dwarf star TRAPPIST-1, which isn't much farther away from us in the cosmic scheme of things—and three of those worlds may be able to support life as we know it. (Proxima

Centauri and TRAPPIST-1 lie about 4.2 light-years and 39 light-years from Earth, respectively. The entire Milky Way is about 100,000 light-years wide.)

So, again: where is everybody? Nobody knows. The Fermi paradox is tougher than a Brazil nut, and scientists haven't cracked it yet. But it's not for lack of trying. They've advanced hundreds of hypotheses to explain it. As varied as these ideas are, they all encompass just a few basic possibilities, as physicist Stephen Webb noted in his book *If the Universe Is Teeming with Aliens, Where Is Everybody?* Let's take a look at each of these three explanation families.

POSSIBILITY 1: WHAT PARADOX? INTELLIGENT ALIENS HAVE ALREADY MESSED WITH US

You may already have wandered off, irritated or incensed that I put the Fermi paradox on equal footing with the beloved crocodile paradox. Perhaps you're now thumbing through a dog-eared copy of *Chariots of the Gods?* or watching YouTube clips of that "alien autopsy" TV special that Fox aired in the 1990s.

Indeed, one possible resolution of the Fermi paradox is that it's no paradox at all, because ET has already journeyed to Earth. Adherents of this explanation often point to UFO sightings and alien abduction stories, topics that you can read about in chapter 10. For our purposes here, suffice it to say that scientists generally don't regard any of these reports as convincing evidence of alien life. (If they did, you definitely would have heard about it.)

There are more subtle possibilities in play as well. For example, what if ET came to our planet long ago, before people were around to be probed? Unless the voyaging aliens were particularly interested in us, this is much more likely than a documented visit, given that our species has existed for just the last 200,000 years of Earth's 4.5-billion-year history and has been capable of capturing encounters on blurry, low-light video for only a few decades.

Let's indulge in some wild speculation, because it's fun! Say Earth has been colonized many times over the eons by greedy, grabby alien civilizations, each of which ground the planet's native species into the dust in the process. (Don't get too high and mighty: pioneering humans have tended to wreak ecological havoc as we've explored the globe.) As astrophysicist and sci-fi author David Brin has pointed out, a history of such oppression could explain why it took intelligent life so long to arise on our planet as well as the radio silence in our galactic neighborhood. Maybe Earth is the only planet for light-years around to have recovered from the ravages of invasion.

If you squint a little, this scenario lines up with the five mass extinctions that scientists have identified in the fossil record. These great purges occurred about 450 million years ago, 375 million years ago, 251 million years ago, 200 million years ago, and, most famously, 66 million years ago, when an asteroid strike wiped out three-quarters of all Earth's species, including the dinosaurs. "It may not be preposterous," Brin wrote in a seminal 1983 paper, to compare the intervals between these extinction events and the time it might take for different waves of invasion to wash over Earth. The dino-killing asteroid could even have been a weapon of war, slung by a space-dwelling alien faction with a beef against their brethren on Earth.

Brin didn't mean to suggest that any of this actually happened, and neither do I. There's no evidence that it did—no spacecraft entombed in ancient amber, no ruins of a 200-million-year-old city—and I certainly wouldn't put any money on it. But it's possible.

POSSIBILITY 2: THEY'RE OUT THERE, BUT WE HAVEN'T FOUND THEM YET

As scientists and other logically minded people often point out, absence of evidence isn't evidence of absence. It's entirely possible that intelligent aliens are (or were) out there, and we just haven't spotted any signs of them yet.

For example, maybe ET hasn't visited Earth because getting here is just too hard. The distances involved in any interstellar trek are mind-boggling. Proxima Centauri is "just" 4.2 light-years from the sun. But that's almost 25 *trillion* miles—equivalent to circling Earth 1 billion times, going to Pluto and back 3,450 times, or jogging around the track at your local high school 100 trillion times. It would take a spacecraft about 75,000 years to get to Proxima Centauri using today's rockets.

There aren't enough honey-roasted peanuts and Sudoku books on Earth to make that trip bearable. Even if we assume that aliens, with their pulsating and extravagantly veined brains, have developed super-fast propulsion tech that puts our puny human gear to shame, there's still a big problem: energy. Say the aliens, like Starfleet engineers, know how to build matter-antimatter engines that can accelerate a ship to 75 percent the speed of light. Just making an Earth–Proxima Centauri round trip with this craft would require 100,000 times more energy than the United States uses in an entire year, physicist Lawrence Krauss wrote in his book *The Physics of Star Trek*. Is the aliens' desire to probe us, or to give the ancient Egyptians some killer pyramid blueprints, really that strong?

Or maybe ET just doesn't want to interfere with the development of life on other worlds—and has hewed to this noble "prime directive" far more successfully than Captain Kirk and his crew have managed to do in the *Star Trek* universe. (Remember when the *Enterprise* gang took it upon themselves to destroy the machine-god Vaal in an original series episode? Vaal seemed like a jerk, but still.) It's even possible that aliens are watching us right now, to monitor our technological progress, figure out how we tick, or keep their bratty kids occupied for a few hours.

Some thinkers take such reasoning a step further, suggesting that we and everything else in the observable universe—yes, even love—may be part of a simulation run on a very fancy alien computer. Before laughing this off, consider how much cooler *Fortnite* is than Burger Time. Those two games were released just 35 years apart, and the hypothetical aliens have had billions of years to come up with amazing graphics and compelling yet believable storylines. Indeed, philosopher Nick Bostrom has argued that the odds we're trapped in a *Matrix*-style pseudo-existence are actually quite high—provided there are a decent number of super-advanced civilizations out there and at least some of them are keen to create convincing virtual worlds, for fun or profit. Given these two assumptions, the number of artificially created universes, or patches of universe, will far outstrip the number of real ones, according to this line of thinking.

Along similar lines, perhaps ET's technical mastery has driven its focus away from the real world and into the virtual, sapping its desire to explore the cosmos or meet any potential neighbors. (Humanity may well succumb to this fate when high-quality virtual reality porn hits the marketplace.)

There are other reasons why advanced aliens may be keeping their heads down as well. Self-preservation springs to mind: what if they're trying to avoid being destroyed or enslaved by big-time cosmic jerks, like the Borg from *Star Trek* or the Galactic Empire in *Star Wars*? Scientists have even suggested that evil aliens may have sent fleets of intelligent, self-replicating "berserker" probes out into the galaxy to hunt for radio transmissions and other signs of intelligent life—and to exterminate any civilizations they find.

Extinction is another possibility. Maybe those berserkers have done a lot of exterminating over the eons. Or perhaps alien civilizations tend to off themselves in relatively short order. Humanity has come perilously close to a nuclear holocaust several times, after all, and we've recently spurred a global mass extinction that may end up claiming our species as well. And yet, with all that, we've been capable of sending signals to other stars for only a century or so.

If a 100-year messaging life span is the norm for civilizations, "then it's as if there are two fireflies that each flick on once during the course of a long night," said Douglas Vakoch, president of METI International, a

San Francisco–based nonprofit dedicated to astrobiology and SETI research. (METI stands for messaging extraterrestrial intelligence—the controversial notion that humanity should reach out to potential alien civilizations, rather than just passively listen.)

The odds that these cosmic fireflies will flash at the same time are, of course, not good. That's sad for them, and sad for any giant space monsters that want to catch them and put them in jars.

It's also possible that ET is trying to get our attention, and we just haven't noticed yet. After all, humanity has been searching for alien transmissions for less than 60 years—the last 0.000001 percent of Earth's history—and always on a shoestring budget.

How shoestring? Well, the US government hasn't bankrolled a SETI operation for a quarter-century. NASA began an ambitious observing project in 1992 but had to stop a year later when Congress cut off the money. (The leader of the defunding push, Nevada senator Richard Bryan, painted the SETI effort as a Mars safari for some reason. "The Great Martian Chase may finally come to an end," Bryan said in 1993. "As of today, millions have been spent and we have yet to bag a single little green fellow. Not a single Martian has said, 'Take me to your leader,' and not a single flying saucer has applied for FAA approval.")

The SETI Institute and other such groups generally rely on private donations to keep the lights on and the telescopes listening. These donations don't always come through. The SETI Institute had to idle its main ear to the universe, the forty-two-dish Allen Telescope Array in Northern California, for four months in 2011, and the original plan called for the ATA to consist of 350 telescopes, but there hasn't been enough cash to complete the build.

Given this situation and the huge scale of the Milky Way galaxy, scientists have not yet been able to mount a comprehensive SETI survey. They haven't even come close.

Tarter often relies on an analogy to get this point across: imagine that you're searching for fish across the entirety of Earth's oceans, and you wade into the surf and scoop up a single glass of seawater. "If you did that experiment and your glass didn't contain a fish, you probably would not conclude that there aren't any fish," Tarter said. "Well, numerically, the amount of searching that we've done versus the amount that we might have to do is equivalent to that one glass of ocean."

We may not even be looking for the right kinds of signals. The SETI search to date has focused heavily on radio waves and to a lesser extent laser-light pulses, because those are technologies that humanity has mastered. But we're already weaning ourselves off radio-wave transmission just a century after inventing it; when's the last time you sharpened your TV's picture by crumpling some tinfoil onto rabbit ears? Would a billion-year-old alien civilization really still be communicating like this, or in any way we could understand? Maybe ET sends messages via neutrinos, the bizarre and unfathomably numerous particles that zoom through planets unimpeded like subatomic Houdinis. (Trillions of solar neutrinos passed through your body in the time it took to read that last sentence.) Maybe the aliens are telepathic. Who knows?

Our current strategy may be akin to trying to eavesdrop on people via walkie-talkie, according to astrobiologist Dirk Schulze-Makuch, who's a professor at the Technical University of Berlin in Germany and an adjunct professor at Arizona State University and Washington State University.

"You probably won't get anything, because everyone is on Facebook," Schulze-Makuch said.

As this discussion shows, many of the ideas bandied about to explain Fermi's paradox basically amount to ET psychology. And that's not the most promising path for a breakthrough: getting inside the heads of super-advanced aliens is beyond us, at least until we stop devoting most of our creative energies to meme generation. (Thank you for indulging this "get off my lawn" moment.)

POSSIBILITY 3: WE ARE ALONE

The last alternative is the most depressing: the cosmic silence speaks volumes.

Maybe Earth is the only inhabited world in the entire galaxy. God loves us that much! Or, if you want to get all science-y about it, the jump from complex organic chemicals to wriggling microbe may be so improbable that it occurred just once, and we hit the jackpot.

This could be a stretch, given how quickly life got a foothold on Earth. Microbes were here by at least

3.8 billion years ago, and perhaps even earlier; some evidence pushes life's emergence back to 4.1 billion years ago, pretty much as soon as Earth had cooled down enough to be habitable. But even if microbes are common throughout the cosmos, intelligent life could still be vanishingly rare. (Astronomers and astrobiologists do actually crack the obligatory joke from time to time: "Hey, we're still searching for intelligent life on Earth!" or "You won't find it on Capitol Hill!") Why? Well, maybe not many planets can offer the long-term TLC required for complexity and smarts to evolve. For example, Earth boasts a large moon that stabilizes its tilt (and thus its climate), and it enjoys the protection of a giant outer planet (Jupiter) whose powerful gravity nudges some dangerous comets away. Perhaps such characteristics are rare for rocky worlds in the habitable zone.

Also, forget what those cartoons showing apes marching toward a proud, pants-wearing future may suggest; there's no "arrow of progress" inherent in evolution. Natural selection favors whatever works, so if simple is successful, simple stays simple. Indeed, that was the story for most of Earth's history. Multicellular organisms don't show up in the fossil record until nearly 600 million years ago—meaning single-celled microbes had the planet to themselves for at least 3 billion years. And there

was another long gap before super-smart animals—
modern humans—came along.

So it might take a really special set of circumstances
to jolt life out of its simple, slimy origins and eventually
reach the point where it can invent radio transmitters,
spaceships, wheely shoes, and other cool stuff. After all,
Earth might still have reptilian overlords if not for that
asteroid strike 66 million years ago, which allowed our
mammalian ancestors to scurry out from the shadows.

There are some other important things to keep in
mind as well. For example, not all intelligence is the
same, as the diversity of life on Earth clearly shows.
Chimps, ravens, dolphins, sea otters, octopi, and a number of other species are smart enough to use tools, but
only humans have built radio transmitters, spaceships,
and wheely shoes. (As far as we know. But if chimps had
wheely shoes, you'd think Jane Goodall would've said
something.) We can't assume that every intelligent alien
species would be technologically smart or able to communicate with us.

The circumstances of their birth may cut many smart
aliens off from the rest of the universe. If our own solar
system is any guide, the most common life-supporting
worlds in the galaxy may be frigid moons and planets
with liquid-water oceans beneath their icy shells—places
like Saturn's moon Enceladus and the Jupiter satellite

Europa. If complex, intelligent life has evolved in such environments—and that's far from a sure thing, given the likely dearth of energy in those dark depths—we might never hear from it.

"How long would it take sentient beings, confined to their pitch-dark liquid habitat by a solid sky hundreds of kilometers thick, to discover that there was a vast universe beyond their world's apparently impenetrable roof?" theoretical physicist Paul Davies wrote in *The Eerie Silence*, his 2010 book about the Fermi paradox. "It is hard to imagine that they would ever 'break out' of their ice prison and beam radio messages across space."

GETTING AN ANSWER

You've made it through the Fermi Paradox Hypothesis Sampler Platter! Did any of the ideas jump out at you? Perhaps the berserkers, for violence and action, or the buried-ocean dwellers, for poignancy? (I picture sallow, eyeless mercreatures sadly strumming lutes.) If so, that's fine, but you probably shouldn't get too attached. We just don't have enough information at the moment to know what's actually going on.

"I find it silly that so many people leap to shout, 'Aha!

I know the answer!'" Brin said. "All we can do is catalog them for now, and maybe rank a 'Top Ten.'"

But we could start getting at that answer, and soon. Say scientists discover a "second genesis" of microbes— tiny organisms completely unrelated to any kind of life as we know it—on Mars, Enceladus, or another solar system body. We would then know that life is not a super-lucky one-off affair, and we'd strongly suspect that it's widespread throughout the galaxy. This news, combined with a continued SETI silence, would also be troubling for anyone who cares about humanity's future, for it would suggest that the bottleneck limiting the number of intelligent civilizations still lies ahead of us. (There could be a benefit, though: if we think we're the only technologically smart creatures in the galaxy, the result-ing sense of responsibility may provide the nudge we need not to destroy ourselves.)

By the same reasoning, getting even a single SETI ping would be a real pick-me-up.

"The detection of a signal—even a cosmic dial tone, with no information—goes on to tell us that we can have a long future," Tarter said. "If somebody else made it through, we can, too."

Chapter 2

Are We All Martians?

Men really may be from Mars—and women, too—if you go far enough back in time.

Yes, this is a shamefully weak and dated joke, but let me explain. The Red Planet wasn't always as cold and dry as a frozen saltine. Before the sun stripped away most of its atmosphere—which happened by about 3.7 billion years ago, after Mars lost its global magnetic field—the planet had lakes and rivers that would have been great for tubing. (Alien-world bonus: exotic vistas and no horseflies!)

Take Gale Crater, the 96-mile-wide hole in the ground that NASA's Curiosity rover is exploring. Not long after the six-wheeled robot's August 2012 landing, Curiosity team members made an announcement that thrilled tubers all over the world: Gale once hosted a lake-and-stream system for at least millions of years at a time. And, mission scientists added, this water was so clean and clear that you could have drunk it. (With no ill effect, that is. You can drink paint or Mountain Dew, if you don't care about consequences.)

It's easy to imagine life taking root inside Gale, or somewhere else on Mars, around 4 billion years ago. In fact, some researchers even think the ancient Red Planet was a better cradle than Earth, partly because our planet was likely covered by a global ocean back then—a real-life Waterworld.

"Now, wait just a minute," you might object. "Earth life needs water, so how could there be too much of it?" Good question! Well, it turns out that water isn't all ice pops and sprinkler showers on a hot summer day. It breaks apart many of the molecules that cells depend on—including nucleobases, key building blocks of DNA and RNA. You know what DNA is; RNA helps "translate" genes into proteins inside the cells of every organism on our planet, even platypuses. Indeed, the trillions of cells that constitute you and me are fixing water damage 24 hours a day, like an army of tiny, super-industrious drywall contractors.

"That's just fine if you've got an advanced organism which is able to evolve a repair system," said Steven Benner, a biochemist at the Foundation for Applied Molecular Evolution in Florida who studies the origin of life. "But when you're trying to get life started, the fact that bases fall apart in water doesn't exactly help you."

RNA is especially susceptible to such damage, he added. That's an even bigger deal than you may realize, because many scientists think RNA was the first genetic molecule ever used by life. Why? Because RNA is as versatile as Meryl Streep, capable of carrying information, copying itself, and sparking a wide range of chemical reactions.

Which brings us back to the Red Planet. Ancient Mars was like good iris-growing soil: wet, but not *too* wet. Though a huge ocean may once have sloshed across much of its northern hemisphere, the planet was never completely flooded, so there were probably lots of places where RNA could have gotten a foothold, the idea goes.

Benner cited another factor that may have worked in Mars's favor as well: the tar paradox. This may sound like a minor *Hitchhiker's Guide to the Galaxy* character, but it actually refers to the tendency of organic molecules—the carbon-containing building blocks of life—to degrade into black sludge when they're zapped with energy. (Think about that time you left the risotto on the stove too long.) Why is this a paradox? Well, simple organics generally don't link together into the big, complex molecules used by life without such risotto-ruining energy influxes.

There are ways around this issue, of course, or you never would have existed. Like a great anti-smoking ad, some substances can halt tar formation in its tracks—notably, minerals that contain boron. Work by Benner and others suggests that boron minerals were much more common on early Mars than on the ancient Earth. And back then, Earth's boron compounds were dissolved and diluted in the global ocean, reducing the odds that small organics with big dreams would encounter them.

Ancient Mars also had more going for it than dry land and boron. Mars is just 10 percent as massive as Earth—meaning the Red Planet cooled down a lot faster after its birth and therefore had temperatures suitable for life before ours did. The disparity in cooling time was increased by circumstance: Earth apparently got walloped by a Mars-size world, dubbed Theia, shortly after the solar system's birth, transforming our planet's surface into a hellscape of roiling magma. (Scientists think the material blasted into space by this cosmic clobbering coalesced to form our moon.) In addition, some studies suggest that phosphates, vital components of DNA and cell membranes, were more common on early Mars than on early Earth.

INVASION OF EARTH?

That's the nutshell case for the Mars-first idea. If it is indeed on the money, then life on Earth may have started with an alien invasion.

We're not talking little green men here, or those weird, bulbous-brained, skull-faced monstrosities from *Mars Attacks!* (They were little but not green, as I recall.) The invaders would have been microbes, and their spaceships chunks of Mars rock blasted off the planet by powerful cosmic impacts.

This scenario is not as far-fetched as it may sound, because pieces of the Red Planet commonly make it to Earth. To date, scientists have identified about 150 Mars meteorites, whose otherness is betrayed by their chemical signatures. Many more are doubtless lying undiscovered at the bottom of the sea, in the middle of the jungle, and in Farmer John's alfalfa field. Some studies estimate that up to 5 percent of all rocks launched off Mars by a violent impact—representing perhaps billions of tons of material over the eons—eventually make their way to Earth.

Intriguingly, the red-rock rain was probably most intense around 4 billion years ago, when life was just getting its start.

Back then, many more space rocks slammed into

Mars (and Earth, Mercury, Venus, and the moon), in a grand, prolonged smashup called the Late Heavy Bombardment. These impactors were building blocks left over from our solar system's planet-formation period and/or comets and asteroids flung inward by fitful "migrations" of the giant planets. (Like new houses, Jupiter, Saturn, Uranus, and Neptune gradually settled. Their orbits are stable now.)

I sense another objection bubbling up: "Interesting theory, but how could Mars microbes survive all the way to Earth? Isn't space deadly?"

Yes, space is uncomfortably cold (or hot, depending upon where you are), radiation-blasted, and oxygen-free. You would not like it. But many microorganisms are much tougher than you (no offense). Tardigrades, adorable little eight-legged beasts also known as water bears, can survive temperatures as high as 300°F and as low as −458°F; they're also not terribly bothered by dehydration, radiation, or enormous pressures that would crush a person to a greasy pulp. *Deinococcus radiodurans*, which scientists have dubbed "Conan the Bacterium," is also a jack-of-all-trades badass, with the signature ability to shrug off radiation doses at least 1,000 times higher than those that would kill you or me in the time it takes to heat up a Pop-Tart.

Both of these little guys have proved their mettle in

space. A few super-hardy tardigrades survived a 10-day orbital trip back in 2007, naked and exposed on the outside of a Russian capsule (though many of their brethren died). Nearly a decade later, researchers stuck a colony of *D. radiodurans* outside the International Space Station for a full year. The poor suckers in the very top layers didn't make it, but their bodies protected the lucky microbes below, who endured the orbital ordeal just fine.

This battlefield-style corpse shielding would aid journeying Red Planet microbes as well. And they'd also get protection from the rock that they're riding on, which could allow some of them to survive the fiery dive through Earth's atmosphere.

This planet-hopping scenario is an example of a broader idea known as panspermia, which isn't nearly as gross as it may sound. (The word derives from the Greek for "seeds everywhere.") Panspermia advocates— panspermians? panspermites?—think that life has spread throughout the solar system, and maybe even the galaxy and the universe, by hitching rides on planet chunks, asteroids, comets, or bits of dust propelled by starlight.

Panspermian Richard Hoover puts his money on comets as life's ride. These icy wanderers are already known to carry amino acids and other organics and could, he said, even pick up living organisms—for example, by flying through plumes of water vapor like the one emanating from the south pole of Enceladus. Comets blast out jets of gas and dust when they get close to the sun and heat up, so they could have delivered life to Earth without even crashing into our planet, said Hoover, an astrobiologist who worked at NASA's Marshall Space Flight Center in Alabama from 1966 until his 2011 retirement.

Most of our solar system's comets live way out in the Oort Cloud, a deep-freeze zone so distant that it makes Pluto and Earth seem like Minneapolis and St. Paul. (The Oort Cloud begins about 5,000 astronomical units from the sun. One AU is the average Earth-to-sun distance—about 93 million miles.) Assuming alien comets are similarly far-flung, it's easy to imagine the sun and other stars swapping comets when they zoom close enough to each other, according to Hoover.

"Life on Earth may not be just extraterrestrial," he said. "It may be from an entirely different solar system."

Why do Hoover and other panspermians think this way? Well, they note the rapid establishment and profusion of life on Earth. Microbes apparently hit the ground running here, quickly diversifying and showing off superfancy tricks like photosynthesis, which may date all the way back to 3.5 billion years ago. (There's considerable uncertainty and debate about the timing of such early milestones, however. All scientists can say for sure about the timing of photosynthesis is that Earth had lots of oxygen in its atmosphere, courtesy of photosynthetic microbes, by at least 2.4 billion years ago.)

"The idea that everything evolved from molecules to bacteria in that period of time strikes me as a little nuts. It's much simpler to say, 'Oh, it just kind of rained down,'" said Gary Ruvkun, a molecular biologist and geneticist at Harvard Medical School and Massachusetts General Hospital.

And alien microbes could make the trip from a nearby solar system in just a few million years aboard their handy comet cabs, Hoover said.

"A few million years is no problem at all for biology,"

he added. "We have microorganisms in my freezer downstairs that are 8 million years old and are still alive."

Ruvkun, on the other hand, is a devotee of "directed panspermia," which was posited in 1973 by molecular biologist Francis Crick—the co-discoverer of DNA's double-helix structure—and biochemist Leslie Orgel. Directed panspermia holds that intelligent aliens have dispersed the seeds of life, either by accident (which would be the most serious littering offense ever, way worse than sneaking away after your dog drops a deuce on someone's lawn) or to satisfy a Darwinian desire to populate the cosmos. Think of the beginning of the *Alien* prequel *Prometheus*, when that giant albino muscleman drinks dissolving juice and spills his inky guts over a waterfall.

Ruvkun reckons the planet-seeding was intentional, but he has in mind a less painful and disgusting—and far more efficient—mode of transportation. He thinks that advanced civilizations have likely sent microbes to colonize other worlds. After all, some folks here on Earth are talking about launching microbes to Mars in the not-too-distant future, to modify the Red Planet's atmosphere and climate to pave the way for human settlement.

Panspermia isn't a crackpot idea, but its adherents are definitely in the minority. All of the above reasoning and speculation notwithstanding, there's no solid evidence for it, so most scientists tend to back the simplest explanation: life started here on Earth. (Researchers also often point out that panspermia doesn't solve the problem of *how* life originated; it simply pushes that origin off in space and time.) Exactly where this may have occurred is a topic of much debate; some researchers back the "warm little pool" cited by Charles Darwin, while others prefer the hot-water seafloor vents known as black smokers and white smokers. The vent idea implies, of course, that the water problem discussed above isn't much of a problem at all.

WOULD WE EVER KNOW?

But say the panspermians are right, and life actually came to Earth from somewhere else. Would we ever know?

Sadly, perhaps not. For example, it would be pretty much impossible to tease out the *Prometheus* liquefied-guts scenario, unless the aliens left something behind in addition to their disgustingly delivered DNA—a spaceship encased in ice, perhaps, or a pamphlet, wedged beneath a very old rock, that begins "So, you're pondering your origins."

We could make some more headway if we found living critters on Mars, Enceladus, or another world within the solar system. Analysis of the aliens' genetic material—we couldn't assume it would be DNA—would quickly reveal whether or not they were related to us. Things would be trickier, though, if all we turned up were signs of life that went extinct billions of years ago; in that case, the molecules might be too degraded to study in depth. After all, we're still waiting for scientists to make a real-life Jurassic Park, and dinosaurs—the ones that didn't give rise to birds, anyway—died out a mere 66 million years ago.

But even figuring out that microbes on Mars and Earth were kissing cousins wouldn't tell us where the family comes from. Did their ancestors zoom into the solar system from afar, riding a comet through the dark and frigid wastes for millions of years? Or did those ancient pioneers get their start closer to home?

"It will be very difficult to say conclusively which scenario is correct and which world in our solar system might have been the starting point," said astrobiologist Chris McKay, of NASA's Ames Research Center in Moffett Field, California.

Scientists might lean away from Earth as the home world in such a case, based on orbital dynamics. Because Earth lies closer to the sun than Mars does, our planet has sucked up many more Red Planet rocks over the eons than the other way around—perhaps 100 times more, in fact. But, as McKay noted, figuring out what actually happened 4 billion years ago would be a tall order.

And what if we determined that the Mars microbes weren't related to Earth life—that they represented a second genesis? Well, that would be a truly huge development. If life arose independently twice just in our little cosmic backwater, it must be everywhere.

"If you can get to two, you can get to a billion," McKay said.

Chapter 3

What Does ET Look Like?

Picture an alien in your head. I don't mean conjure up a skull-dwelling parasite that feeds off your bad dreams and insecurities, though this is indeed an option. Any alien, anywhere. What do you see? I'm willing to bet it's a smallish humanoid with a bald, oversize head, creepily slender fingers, and big, dark, gorgeous eyes that a thousand astronauts could drown in.

You'll be pleased (or terrified) to hear that this creature, the archetypal "Grey" from abduction stories and sci-fi, does indeed skulk around out there in the vastness of space—provided our universe is infinitely large. In an infinite universe, every combination of atoms and molecules that's physically possible will manifest somewhere—an infinite number of places, in fact, as long as matter is spread more or less evenly throughout the cosmos, as seems to be the case. Infinity is weird like that. So, in this scenario, the Greys are just one attraction in an endless cosmic menagerie studded with fantastic beasts, from real-life versions of Yoda and Dr. Zoidberg to doppelgangers of you, me, Al Capone, Usain Bolt, and every other person who has ever trod the Earth. (Yes, you are a fantastic beast! Don't let anyone convince you otherwise.)

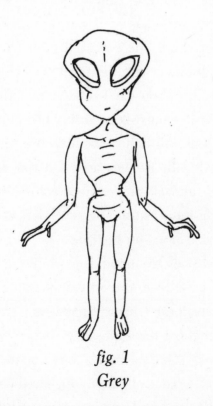

fig. 1
Grey

Unsettling, right? You thought you were special. Well, the good news is that you'll almost certainly never cross paths with any of those weird alternate yous out there. The closest one is about 10 to the power of 10^{29} meters away from you right now, according to calculations by physicist Max Tegmark. As the weirdness of that number suggests, this is a mind-bendingly huge distance—far bigger than the entire observable universe, an ever-expanding sphere that's currently about 93 billion light-years across.

A quick note here: the extent of the observable universe is the distance that light has traveled since the beginning of time. You can't observe anything without light, after all. It seems like that bubble should have a radius of 13.8 billion light-years, because the Big Bang happened 13.8 billion years ago. But space-time itself is expanding—and at an ever-accelerating rate, thanks to a mysterious force astronomers call dark energy—and light gets carried along for the ride.

But let's get back to that all-important question: does the universe actually go on forever, like an elementary-school play? It doesn't have to, after all. Cosmologists generally recognize three possible shapes for our universe: spherical, also known as closed; open, like a saddle; or flat, like a piece of paper. Open and flat universes are infinite. But a closed one is not; you could, theoretically at least, go Space Magellan on it, starshipping all the way around the sphere and ending up back where you started.

Astronomers have figured out how to measure the shape of space-time on a grand scale—by scrutinizing the cosmic microwave background, the ancient radiation left over from the Big Bang. In a closed universe, CMB light would bend in a certain subtle way, while it would arc slightly differently in an open universe. If the cosmos were flat, there would be no curve to this light at all.

And that's just what the measurements by NASA's Wilkinson Microwave Anisotropy Probe and the European Space Agency's Planck satellite have shown: zero curve. There's a little wiggle room in there, however. Planck's data, the most precise available, have a margin of error of plus or minus 0.4 percent. So if the cosmos isn't infinite, it's still incredibly large—250 times wider than the observable universe at a minimum, according to one 2011 estimate.

You may be wondering just what the universe is expanding *into*. This is a reasonable question, but the answer is pretty unreasonable, at least to our poor little ape brains. Physicists don't think anything lies beyond the universe. There's no dark realm of broken dreams, no clean white space filled with exalted souls, not even nothing. The universe is just getting bigger, without displacing or diminishing anything else.

Flatness is also predicted by the theory of cosmic inflation, the best-accepted model of the universe's first few moments of existence. According to this idea, space-time expanded far faster than the speed of light for a brief stretch (pun intended), beginning about 10^{-35} seconds—one trillionth of a trillionth of a trillionth of a second—after the Big Bang. This wouldn't violate Einstein's theory of special relativity, by the way: nothing can move faster than light through space, but inflation was an expansion of space-time itself. Most versions of inflation theory also posit that the infant cosmos pinched off into multiple expando-bubbles, each of which grew into a separate universe. If this is true, then we live in one of many parallel universes that together compose a multiverse. Some of these other universes are bound to be very weird, perhaps with different physical laws and a different number of dimensions than the four we're used to. (Remember: time is a dimension.) And you definitely wouldn't have to worry about running into your parallel-universe twin.

"These other domains are more than infinitely far away, in the sense that you would never get there even if you traveled at the speed of light forever," Tegmark wrote in a 2004 paper. The reason, he added, is that the domain between our universe "and its neighbors is still undergoing inflation, which keeps stretching it out and creating more volume faster than you can travel through it."

The number of these inflationary parallel universes may even be boundless—meaning we might be able to get to infinity even if our own cosmos looks like an enormous beach ball.

There are other possible types of parallel universe as well, as Tegmark has explained. For example, there's the quantum-mechanics parallel universe, which famously holds that every possible measurement outcome is real in the macroscopic world just as it is in the bizarre, incomprehensible world of the very small. This Hitler-actually-won-World-War-II parallel universe was first formally postulated in the 1950s by physicist Hugh Everett, the father of Eels frontman Mark Oliver Everett.

So, while we don't know the universe's size and shape for sure, theory and observations are both pointing in the same direction—a cosmos in which, somewhere, a Zoidberg and an alternate you help slather moisturizer onto the back of a haughty Grey starship captain prepping for a colonization mission.

If the Grey captain deigned to let you come along for the ride, you might see something far weirder than him or the Zoidberg. The starship could conceivably

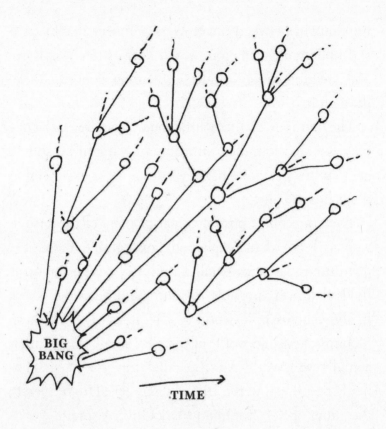

cruise past sentient clouds of interstellar gas and dust, and creatures that live on super-dense stellar corpses known as neutron stars, provided these things are indeed possible.

What a fascinating universe! Aren't you glad you live here?

MEETING THE NEIGHBORS

"Everything that's physically possible" is not a particularly satisfying answer to the question that led off this chapter. So let's ask something slightly different: if extraterrestrial life does indeed exist, what are some of its most abundant forms? What sorts of creatures are likely to be our neighbors?

Sorry, sci-fi fans: they probably won't be *too* alien.

At its core, life as we know it is just elaborate and highly organized chemistry—a fact recognized by NASA's working definition of the term, which the agency adopted in the mid-1990s after much debate and discussion: "A self-sustaining chemical system capable of Darwinian evolution." (Note to creationists: life and evolution are as inseparable as Goldie Hawn and Kurt Russell.)

Here on Earth, that system centers on carbon and liquid water. The molecules that life is based on—DNA and proteins, for example—are long strings of carbon atoms with a variety of chemical groups branching off on the sides. And water is like Amazon for cells, delivering salts, nutrients, and everything else they need for living (and also carrying away their waste, which fleets of Amazon drones will probably do for us before too much longer). The liquid part is key here: it would be tough for cells to get ahold of stuff in a gaseous medium, where atoms and molecules would be dispersed far and wide, or in a solid, where they'd be locked in place. (Rocks and bricks don't flow very well.)

This setup makes a lot of sense. Carbon is a fantastically fertile and flexible atom: it is capable of bonding with four other substances at the same time, allowing the construction of big, complex, and stable molecules. Water is great at dissolving (and therefore delivering) stuff; chemists don't call it the universal solvent for nothing. And H_2O happens to be a liquid at the temperatures that Mother Nature dictated Earth would experience, thanks to our planet's distance from the sun and atmospheric composition.

Not surprisingly, the two work well together.

"If you want complex chemistry in water, carbon really is quite, I would say, unique," said astrobiologist and chemist William Bains, a visiting researcher at MIT. "It is very unusual in the amount of complex chemistry it can sustain in water."

Carbon and water are also both cosmically common. H_2O is the most abundant solvent in the universe; it's found pretty much everywhere, from interstellar clouds of dust and gas to the comets, asteroids, planets, and moons of our own solar system. The galaxy's naturally occurring carbon comes in stunningly complex forms. For example, astronomers have found dozens of amino acids in meteorites, meaning that the building blocks of life as we know it commonly arise of their own accord in deep space. Many researchers therefore think long-ago comet and asteroid impacts were crucial to the rise of life on Earth, delivering these building blocks, along with lots and lots of water, to our planet.

Add all this up, and the universe seems primed for carbon-water life. And, importantly, we know this sort of biochemistry works; Earth is living proof. I'd guess that most aliens we eventually stumble across practice carbon-water chemistry, or at least started out doing so. This doesn't necessarily mean that ET will use DNA as its genetic material, or rely on the same 20 amino acids that Earth life does. Other carbon molecules could conceivably get the relevant jobs done, and it might actually be quite surprising if two different life systems came up with the exact same solutions—unless, of course, panspermia is a thing and we're all related.

Now, I'm not suggesting that alternative biochemistries can't or don't exist. Indeed, carbon-based life could very well use a solvent other than water—for example, ammonia, which is also pretty common throughout the universe and plays well with carbon. (Not with *your* carbon, though, unless you like skin burns, convulsions, and liver damage.) Carbon doesn't have to be a major player, either. There are good reasons why sci-fi keeps trotting out silicon-based organisms, like the stony Horta from the original *Star Trek* TV series. Silicon lies right below carbon on the periodic table, meaning the two elements are pretty similar chemically. For example, just like carbon, silicon can make four different bonds simultaneously. Silicon isn't a great scaffold for life on Earth—it reacts too easily with water and oxygen, forming sand and rocks—but the stuff could be a viable option on worlds very different than our own.

One possibility is Saturn's huge, haze-shrouded moon, Titan, which is bigger than the planet Mercury and sports lakes and seas of liquid hydrocarbons on its frigid surface. Some astronomers think silicon could be the atom of life in these otherworldly seas, or in pockets of liquid nitrogen, which may exist beneath the icy crust of Neptune's largest moon, Triton.

Bains isn't getting his hopes up for a silicon-based life discovery, however. The exotic liquids that could support it tend to be extremely cold—you'd lose your toes if you dipped them into Titan's seas, which clock in at -292°F, and Triton's potential nitrogen habitat would be even chillier. That's a serious problem, because solvents aren't like mailmen; they can't be counted on to deliver when the temperature drops.

"I suspect that the amount of larger molecules you could dissolve in really cold solvents, no matter what they're made of, is just so small that you can't build a sort of functioning biochemistry," Bains said. "It would just be so dilute that it would take geological epochs to do anything."

In other words, unlike a warm little pool here on Earth, a liquid-methane lake may be an organics desert, carrying few big compounds that are free to mix and match and build upon each other.

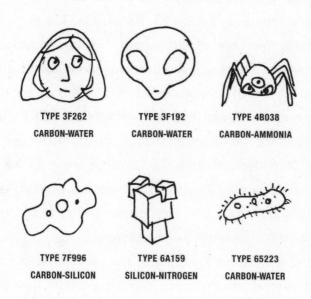

TYPE 3F262
CARBON-WATER

TYPE 3F192
CARBON-WATER

TYPE 4B038
CARBON-AMMONIA

TYPE 7F996
CARBON-SILICON

TYPE 6A159
SILICON-NITROGEN

TYPE 65223
CARBON-WATER

MOSTLY MICROBES

That's a brief and speculative dip into aliens' potential insides. What about that vastly more important consideration: their looks?

Well, you'd probably need a microscope to enter any aliens that may exist in our cosmic neighborhood on Hot or Not, if Earth's history is any guide (and there's no guarantee that it is; extrapolating from a sample size of one is a very dicey business). Recall that life on Earth remained microbe-only for about 3 billion years after first arising. Those selfish little singletons didn't glom together to form animals until 600 million years ago or so, and the planet had to wait another 580 million long years for the first turkeys to evolve. Humans are extreme latecomers: though the hominid line goes back a few million years, our species has been on the scene a mere 200,000 years, and we became truly modern only in 1999, with the launch of LiveJournal.

Don't get too cocky, though: the planet still belongs to microbes if you go by the numbers. There may be 1 trillion species of these little guys, and their total weight tops that of all the animals in the world combined. Then there's the microbiome: no matter how well you wash your hands, your body will always harbor more bacterial cells than human ones. The saving grace is that bacteria are much smaller than native human cells on average, so you're only a tiny percentage bacteria by weight.

So, a scrupulously moisturized Grey captain who visited Earth at some random time during its 4.5-billion-year existence would be 150 times more likely to encounter a microbe-only planet than one with turkeys. Poor captain! Missing out on fleshy, dangling snoods! And he or she would have just a 0.0000004 percent chance of seeing humanity in all of its current meme-sharing glory.

The ET microbes we come across will probably look familiar to us: little membrane-bound bags of important molecules and structures (though the superficial resemblance could of course mask huge differences in biochemistry and metabolism). But if we manage to scrape up anything big enough to pinch or hug or probe or dandle, all bets on appearances are off.

Think about the many amazing forms that life on Earth takes: carnivorous plants that lure insects into a stew of digestive juices with the sweet smell of fake nectar (Raffles' pitcher plant); scaly, limbless creatures that can see heat and swallow prey heavier than themselves whole (snakes—specifically boas, pythons, and pit vipers); giant, armored tank–like beasts with gore-horns sprouting from their stone-gray snouts (rhinos); boneless, sucker-armed, sharp-beaked meat sacks that move via jet propulsion and can change the color and texture of their skin at will (octopi and cuttlefish); and hyperactive, egg-laying, semiaquatic furballs whose duck-like bills can sense the electric fields generated by their prey's twitching muscles (platypus).

That's just the tiniest sampling of what's alive today; the lost menagerie of extinct creatures is vastly larger and quite a bit weirder. Take *Hallucigenia*, which is thought to be an ancestor of today's velvet worms. It was unclear from the first *Hallucigenia* fossils which side of the small and spiky creature was up or where its head was. Musing about whether aliens are likely to have tentacles or tusks, two legs or 10, therefore seems like a fool's errand; would an alien biologist ever conjure up a platypus when dreaming about life on Earth? (Indeed, when the first platypus specimens made their way from Australia to England in the late 18th century, some European biologists thought the animal was a hoax—a stitched-together amalgam of a duck and a mole.)

Still, there are a few things we can say. For starters, the particulars of ET's home world will shape his or her looks. Take Kepler-452b, a potentially life-supporting alien planet that lies 1,400 light-years away from us. Kepler-452b is about 60 percent bigger than Earth and five times more massive, so any creatures that may exist on its surface would have to deal with gravity twice as strong as the force that keeps you and me grounded. It's hard to imagine anything like a giraffe evolving there. Tall, spindly plants would likely be a no-go as well. Rather, most land-dwelling Kepler-452b-ians would probably be short and stout, like badgers or tortoises. (Ocean creatures would have more structural leeway, since water would support their weight.)

But life-forms on Kepler-452b could be familiar in some respects. The leaves of ferns and ficuses here on Earth are green because the main photosynthesizing pigment they contain, chlorophyll, bounces that component of sunlight back like a bad check, preferring instead to use red and blue wavelengths. Assuming Kepler-452b's atmosphere is more or less Earth-like, the planet's sunlight is likely similar to ours; the planet orbits a sun-like star at about the same distance that Earth circles the sun. So if plants (or something like them) live there, they may be green as well.

By that same reasoning, worlds circling stars very unlike the sun could boast seriously weird vegetation. Plants that soak up light from a dim dwarf star, such as TRAPPIST-1 or Proxima Centauri, might not spurn any wavelength in the visible range and therefore appear black to us. And an alien elm looking up at a big, hot star could well have blue leaves, reflecting away energetic wavelengths that could give it a vegetable sunburn if allowed in.[1]

A little knowledge of an alien's home planet or moon also permits us to indulge in some not entirely pointless musing about its habits and lifestyle. For instance, energy is likely in short supply on the frigid surface of Titan, which lies about 10 times farther from the sun than Earth does. So any creatures scratching out a living there may make sloths look positively hummingbird-esque by comparison, barely moving or even breathing over the span of a typical human lifetime. And Titan dwellers may have to endure this stultifying (to us) existence for thousands and thousands of years.

"Our life span is pretty short because there's so much damage [to our DNA] from UV radiation and other radiation. We're constantly under repair. And then, of course, it's warm; we have to act fast and move fast," Schulze-Makuch said. "But if you're in an environment like this, and maybe have a really cold solvent, too—methane or ethane—then you can think about a lot of very exciting, very different possibilities."

---

1   Similar reasoning could explain why plant photosynthesis on Earth shuns green light, which seems odd and inefficient, considering that the sun emits a lot of energy in this wavelength range. Green light may just be too much of a good thing.

Carl Sagan was good at imagining very exciting and different possibilities. In 1976, he and astrophysicist Edwin Salpeter laid out a hypothetical ecosystem high up in Jupiter's skies, full of airborne creatures the duo dubbed "sinkers," "floaters," and "hunters." The little sinkers might be photosynthetic, or they could feed off complex carbon molecules swirling in the atmosphere around them. The postulated floaters were stranger still—beings the size of cities, filled with hydrogen gas to stay aloft. The winged hunters would terrorize the blimp-like floaters, swooping in to nip and tear at their enormous, jellyfish-like bodies.

Telescopes and Jupiter-orbiting spacecraft have seen no evidence of any such beasts, and there are many reasons to think that life in the skies of gas giants is an extreme long shot. (For one thing, these planets' atmospheres are patchy and super-turbulent, and currents would likely carry creatures from potentially habitable areas down to deadly hot high-pressure zones quickly and often.) But that 1976 study is a fun and interesting thought experiment, and my inner 12-year-old is thrilled that the terms *floaters* and *sinkers* now have currency in serious scientific discourse.

INTELLIGENT LIFE: ENTER THE GREYS?

Speaking of thought experiments, have you heard of the dinosauroid? This beast was born a while back in the mind of Canadian paleontologist Dale Russell, who was trying to project what a relatively big-brained dinosaur might have evolved into had Earth dodged that famous asteroid bullet 66 million years ago.

Russell started out with a real-life dinosaur called *Troodon* and the assumption that natural selection generally favors greater intelligence. He ended up with a marvelous creature that could pass for you or me, if we spoke in birdsong and had scaly greenish skin, three-fingered hands, internal genitals, and smooth, nipple-free chests.

If Russell's rendering is right, Earth was pretty much destined to have humanoid overlords, with or without dramatic instances of death from above. Some scientists do indeed think he was on to something—most notably Cambridge University paleontologist Simon Conway Morris, who has long argued that evolution is a predictable thing that repeatedly generates similar outcomes, given similar environmental conditions. As evidence, Conway Morris cites the many examples of convergent evolution found in the fossil record and all around us. For instance, sharks and dolphins developed the same basic body plan, despite being only distantly related. And intelligence has evolved many times across a variety of animal lineages; creatures as diverse as octopi and crows have shown the ability to use tools. (Intelligence is a tough and slippery concept, of course. Humanity tends to define it as the ability to think more or less like us—to learn on the fly, use tools, dominate our neighbors and environment, and so on. This bias, which doesn't take into account other creatures' evolutionary history or ecological niches, may blind us to some cognitive feats in the natural world.)

fig. 2
Dinosauroid

If you're a fan of the Greys, you're probably smiling right now, because this reasoning also applies to alien worlds. If life takes root on an Earth-like exoplanet, Conway Morris has said, it has a pretty good chance of radiating into creatures that look a lot like Earthlings — including us.

"I would argue that in any habitable zone that doesn't boil or freeze, intelligent life is going to emerge, because intelligence is convergent," he said back in 2015, when his book *The Runes of Evolution* came out. "One can say with reasonable confidence that the likelihood of something analogous to a human evolving is really pretty high."

But not everyone thinks evolution is so deterministic. Some folks argue that contingency has been a much bigger player than convergence in the history of life on Earth—in other words, the dino-killing asteroid made a really, really big difference. The late evolutionary biologist Stephen Jay Gould summed up this camp's viewpoint in his book *Wonderful Life*: "Replay the tape a million times from a Burgess beginning, and I doubt that anything like *Homo sapiens* would ever evolve again."

Gould was referring to the Burgess Shale, a famous fossil site in the Canadian Rockies that preserves a profusion of soft-bodied animals from the Cambrian period of Earth's history, more than 500 million years ago. In Gould's view, pretty much any of these ancient weirdos eventually could have emerged as the planet's dominant evolutionary force, given an asteroid strike here or a cataclysmic volcanic eruption there. Nothing was preordained; there wasn't anything particularly special, or inherently better, about the gloopy globs that led to us.

Harvard University evolutionary biologist Jonathan Losos has written a book all about contingency and convergence, called *Improbable Destinies: Fate, Chance, and the Future of Evolution*, which discusses the dinosauroid in depth. He thinks it's unlikely that intelligent aliens would look much like us. Many creatures have indeed converged on similar body plans here on Earth, Losos said, but there are lots of singletons, too—the glorious platypus, for example. If Earth and intelligent humanoids are such a matched pair, why did it take four billion years for us to evolve? Why aren't there other creatures like us scattered throughout the fossil record?

Evolution builds on the forms and functions that have come before, as anyone with a bad back can appreciate. When our ape-like ancestors began walking upright on the African savannah a few million years ago, natural selection wrenched a spine that had evolved for life in the trees 90 degrees, forcing it to support new stresses and loads that it still has a hard time handling.

Starting points matter a lot. The more distantly related two populations or species are, the less likely they seem to be to converge on the same solutions to evolutionary problems, Losos said. (For example, the contingency crowd has a different vision of the dinosauroid, one that takes into account *Troodon*'s lineage. *Troodon* was a theropod, the branch of dinosaurs that gave rise to birds, so the contingent dinosauroid is basically a big, heavy-headed, possibly weapon-wielding crow. An illustration in *Improbable Destinies* shows this badass bird brandishing a spear.)

That's bad news for any alien hunters hoping to bag

some Greys, or any other humanoid exo-creature, according to Losos. "Life on another planet is about as distant as you can get," he said.

Losos doesn't profess to know what intelligent aliens might look like, but he views the weird, vapor-writing heptapods of the 2016 film *Arrival* as a more thoughtful stab than, say, the Greys, or the lizard people of the 1980s TV miniseries *V* (who could be the original dinosauroid's evil cousins), or all those *Star Trek* races that are basically just humans with a new coat of paint.

"My guess is that alien life-forms are going to be very different—so different that we will share very little commonality with them and, as such, have a really hard time perhaps even realizing that they're life, and then trying to find any sort of way of communicating," Losos said.

So the jury is out on how "realistic" the Greys may be. Why, then, do they spring to mind whenever we hear the word *alien*? Part of it certainly has to do with those old-timey, bumpy-faced *Star Trek* creatures and their forerunners. Back before good computer effects, TV and movie aliens were played by actors, and slapping a mask on someone or painting his or her skin blue was a lot cheaper and easier than designing an intricate and imaginative costume. (I don't mean to pick on the original *Star Trek*; some of its aliens, like the Horta, were quite imaginative. But there certainly were a lot of humanoids, a coincidence that a 1993 episode of *Star Trek: The Next Generation* explained by invoking a long-ago directed panspermia campaign.) This was sci-fi's formative period, so humanoids got locked in as alien archetypes.

But there are other reasons as well. For example, we care more about characters that we can identify with. It's hard to get emotionally invested in a sentient cloud of gas and dust, no matter how compelling its backstory may be.

DIGITAL LIFE

Whatever body plan evolution "chooses" for intelligent aliens may not keep them prisoner for long. Some astronomers and astrobiologists think that when we finally make contact with ET, we'll be talking to a machine.

The reasoning goes like this: humanity is just a century into its truly technological phase—the era of radio communication, spaceflight, and cell phones—and we've already invented computers with greater processing power than our own brains. Before too much longer, we will have developed artificial intelligence so strong that we can basically merge with it, uploading ourselves and becoming immortal. (Unless this super-smart AI goes all SkyNet and destroys humanity. Many folks worry about this possibility, including Elon Musk, who has called rogue and unregulated AI the greatest existential threat we face.)

The futurist Ray Kurzweil has predicted that this world- and humanity-changing event will occur around 2045. Seth Shostak, a senior astronomer at the SETI Institute, has stressed that even if Kurzweil is off by a few centuries, any aliens out there will have a vanishingly small window to ask us what we ate for breakfast. And there's pretty much no chance our interstellar pal would wax rhapsodic about alien bacon, because his or her civilization is bound to be much older than ours (assuming our trajectory is typical, of course).

For SETI scientists like Shostak, this is more than just idle speculation: it informs their alien-hunting strategies, as we shall see.

Chapter 4

Do Aliens Have Sex?

Not enough of it, obviously, or they wouldn't have to come all the way here with their dog-eared human anatomy textbooks and lubed-up probes.

Seriously, though: many ET species probably do get busy, if life on Earth is any guide.

"Given how often it has evolved, in a number of different ways and guises, it'd be really surprising to me if there wasn't some genetic mixing on another planet," said biologist Sarah Otto, director of the Biodiversity Research Centre at the University of British Columbia.

Note the term *genetic mixing*, for that's what sex is all about. It's a way for organisms to diversify their descendants' genomes, so little Junior can better deal with

increasing temperatures, new diseases, annoyingly per-
sistent parasites, and everything else that nature may
throw at him or her.[1] Love and pleasure and empower-
ment and revenge are purely secondary considerations,
enticements to get us across the gene-mixing finish
line.

Such genetic flexibility and variability must be
hugely important, given the prevalence of sex—all
birds and mammals get it on, as do almost all reptiles
and amphibians, and the coolest plants (flowers!)—and
the costs associated with it. Think about it: asexual reproduction is easy and uncomplicated, and creatures
that practice it give 100 percent of their genes to their
offspring. The whole point of reproduction—and of
life in general, unless you want to get all metaphysical
about it—is to pass down your genes, and sex, like an
incredibly bold stockbroker, takes a 50 percent cut off
the top.

---

1   This widely accepted idea is known as the Red Queen hypothesis, for
     it invokes a scenario in which sex helps critters keep up with a chang-
     ing environment. As the Red Queen put it in Lewis Carroll's *Through
     the Looking-Glass*: "Now, here, you see, it takes all the running you
     can do, to keep in the same place."

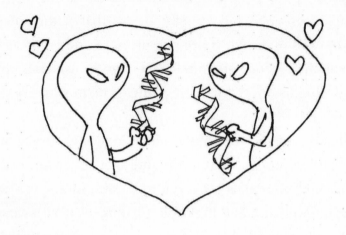

But the gene-mixing game isn't all about sex. While dangly bits make everything more fun, they're not strictly necessary. For example, bacteria don't reproduce sexually—they just make copies of their DNA and then split in two—but these little beasties still manage to get ahold of outsider genes, and in three different ways, no less. First there's conjugation, in which one bacterium transfers some genes (in many cases a weird little circular collection of DNA known as a plasmid) directly into a buddy, often via a specialized bridge-like structure. Then there's transduction, in which viruses spread genetic material from one bug to another. Finally, bacteria sometimes just take in DNA that's floating freely in the environment, sucking it up like strands of tiny spaghetti. The technical term for this is transformation, for some reason, but let's just call it gene slurping.

These strategies may not have evolved expressly to promote gene mixing; that may just be a beneficial side effect for the bacteria. For example, the driving force behind conjugation may be the plasmids' selfish desire to replicate themselves, Otto said. And gene slurping may be mostly about getting a free meal.

It's reasonable to assume that alien microbes do something similar, grabbing snippets of genetic material (again, we can't assume it will be DNA) here and there, either from each other or from the environment. But such a half-assed approach doesn't work well for big creatures, who have lots of cells to get the newly acquired genetic information into, Otto said. That's why we have sex—and why big aliens probably do, too.

"If there are large organisms, they're likely to engage in a more formal way of exchanging genes," Otto said, trotting out a euphemism that wouldn't be out of place in a Kansas public school's abstinence-only pamphlet.

What might a more formal way of exchanging genes look like on an alien planet? Well, it seems to me that pretty much anything could go in ET's boudoir, for two main reasons: (1) infinite (or huge) universe = infinite (or extensive) kinkiness and (2) just look at the crazy sex lives of creatures on Earth! Here's a tiny sampling:

Male bedbugs bypass females' reproductive tracts, injecting sperm right through the poor ladies' body walls using sharp, syringe-like penises. (If you needed another reason to hate bedbugs, I give you *traumatic insemination*, the actual term for this terrible behavior.) Male anglerfish that manage to find mates in the dark ocean depths become permanent sex parasites when they grow up, clamping onto females' bellies with their fearsome fangs and secreting a chemical that fuses the duo's flesh in an ultimate 'til-death-do-you-part. Snakes have elaborately spiky penises—two of them, actually, but they use only one at a time—that latch onto a lady's insides and prevent her from crawling away before the deed is done. And some male animals don't use penises at all during sex; for example, spiders transfer sperm using modified mouthparts called pedipalps.

And all this talk of males and females may be a bit provincial: there's no good reason to assume that an ET species will have two separate sexes. After all, lots of critters here on Earth—including earthworms, most slugs and snails, and the vast majority of plants—are hermaphrodites. At the other end of the spectrum, some fungus species have dozens or even thousands of different sexes. (One quick slug-sex note before we move on: sometimes a banana slug's penis gets stuck in its partner during sex. If the organ cannot be pried free, one of the

two slugs will bite it off, a process known as apophallation. That's right: banana slugs commonly chew off their
own penises during sex. Apophallated penises don't grow
back, but that's cool; mutilated slugs have the fallback of
becoming ladies full-time.)

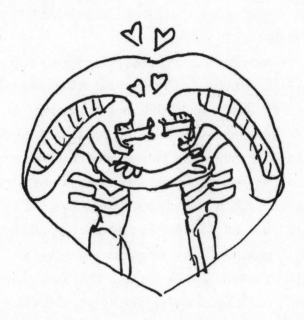

MAKING A FEW GUESSES

So diversity and weirdness will doubtless reign on alien
worlds, just as they do on Earth. But that doesn't mean
we can't make any educated guesses at all about ET's sex
life. For example, it probably takes two to tango—rather

than three, or four, or twelve—on most planets where sex
has evolved, Otto said.

"That's because the costs of sex increase with how
many partners you need to get together," she said. "You
have the waiting time that goes with finding a partner,
the risk of being rejected—all those things go up and up
and up as you have more partners that are needed to engage in that active union."

Don't worry, fetishist weirdoes. This doesn't imply that
alien orgies cannot or do not exist. Otto was talking
about ET's likely sexual needs, not its preferences.

Also, the more mobile ET is, the more likely its species
is to have separate sexes. This is a powerful pattern here
on Earth: 94 percent of plant species are hermaphroditic,
whereas a similarly overwhelming percentage of animals
are not, Otto said. This makes a lot of sense, as most evolutionary patterns do. If you can move around a lot, like a
barn owl or a blue whale, you've got a decent chance of
finding a mate. But if you're literally rooted in place, lov-
ing yourself may be the only option.

It also seems reasonable to speculate that aliens who
live on planets with dramatic climate shifts will generally
rely on genetics to determine the sex of their babies, just
like humans do. You may not have realized that there
are other ways to do this, but nature is full of delightful surprises. In many egg-laying reptiles, for example,

offspring sex depends on incubation temperature. For sea turtles, relatively hot nests produce females, whereas cool conditions generate boys. At a certain in-between temperature, equal numbers of males and females result.

This obviously wouldn't work out too well if your world swung swiftly from balmy to hot. Dudes would die first, a fate they'd probably deserve. But the ladies would soon follow, and there would be nobody around to bury their corpses. Sadly, this scenario could play out for many reptile species here on Earth if our planet continues to warm.

Oh, and one final note: if aliens have indeed dropped by Earth now and again to do some people probing, they've probably been motivated by scientific curiosity rather than base carnal desires. Any creature advanced enough to get here may well have transcended its wetware and gone digital, leaving sex far behind. But if we could somehow figure out how to communicate with such super-intelligent extraterrestrials—a very big *if*—we might be able to dive into their deep-history databases to find out how they used to get busy back in their biological days. And maybe we could ferret out the specs for a really good VR porn system as well.

Chapter 5

What Are We Looking For?

Percival Lowell was a big believer in Martians—
specifically, cotton-mouthed creatures desperate to bring
water from the polar ice caps to the temperate regions
of their drying, dying planet. Between 1895 and 1908,
the American astronomer published three books laying
out the case for the Very Thirsty Martian, which was
based primarily on an extensive network of long, straight
Red Planet canals—nearly 200 of them!—that Lowell
thought he saw through a telescope at his observatory in
Flagstaff, Arizona. (The Lowell Observatory, which was
founded in 1894, remains an important research facility
today.)

In Lowell's eyes, there were other pieces of evidence
as well. For example, in 1908, his Lowell Observatory

colleague Vesto Slipher reported spotting water vapor in the Martian atmosphere, suggesting that the planet might indeed be able to support little green civil engineers—for a little while longer, at least. Though Slipher was almost certainly mistaken—Mars's thin atmosphere is more than 95 percent carbon dioxide and contains only tiny traces of water—he still had the coolest name in 20th-century science, and nobody would ever take that away from him. (Philo Farnsworth, who invented the all-electronic television, is a distant second.)

Lowell was wrong, of course. There is no planet-wide irrigation system on Mars. Many astronomers argued this at the time, and it finally became firmly established in 1965, when NASA's Mariner 4 spacecraft beamed home up-close images of a surface rich in craters but sadly lacking in canals and desiccated corpses.

This little tale doesn't mean that looking for alien engineering projects is weird or stupid or pointless. Indeed, researchers still keep their eyes open for signs of giant space structures—and they got tantalizing hints of a hit just a few years back, as we shall see. But for the most part, astrobiologists are thinking much smaller than Lowell did, because odds are that microbes and other relatively simple life-forms far outnumber big and brainy beasts throughout the galaxy. That's especially true in our own solar system. We've managed to explore it a bit, and our spacecraft and telescopes have spotted no homes hewn, Petra-style, into Mars's red buttes, no Sinker City gleaming porcelain-white amid Jupiter's cloud tops.

ALIEN GAS

There are a few different ways to bag an alien microbe. For starters, you can catch the little fellow doing his business, emitting gases out into the air or soil of his home world. Scientists have already taken this strategy into the field, most famously with NASA's twin Viking 1 and Viking 2 landers, which touched down on Mars six weeks apart in 1976.

Both craft carried three life-detection experiments, two of which aimed to spot specially tagged carbon in gases given off by living microbes. The Vikings picked up intriguing chemical signals in the Red Planet's dirt, and some folks argue to this day that the landers discovered evidence of Martian life, as we shall see.

Then there's the European-Russian ExoMars Trace Gas Orbiter (TGO), which arrived at Mars in October 2016 and began sniffing for methane in the planet's skies 18 months later. Methane gets astrobiologists excited because it's a potential biosignature; more than 90 percent of the methane in Earth's atmosphere is generated by life. (Yes, cow farts famously contribute, but free-living microbes are the primary producers here.) And a Mars methane mystery has been brewing for the past 15 years or so, since Earth-based telescopes have flagged apparent wisps in the Red Planet's skies on several occasions, and NASA's Curiosity rover rolled through a plume of the stuff in late 2013 and early 2014. Excitingly, these emissions have to be recent, because ultraviolet radiation from the sun breaks down methane molecules within a few hundred years of their appearance in the Martian air. (Unlike Earth, the Red Planet doesn't have a UV-shielding ozone layer.) Researchers hope TGO helps them nail down the quantity of Red Planet methane, and where exactly it's coming from.

It's possible to expand biosignature searches beyond our own solar system as well—and alien hunters will start doing so very soon, if all goes according to plan. NASA's $8.8 billion James Webb Space Telescope (JWST), the agency's much-anticipated successor to the famous Hubble Space Telescope, should be able to scan the atmospheres of some nearby exoplanets after it launches in 2021. And a few years later, a whole phalanx of planet sniffers will join JWST in the fray—the

European Extremely Large Telescope, the Giant Magellan Telescope, and the Thirty Meter Telescope, all of which are ground-based instruments. NASA is also developing another spacecraft that could do such work—the Wide Field Infrared Survey Telescope, which has been targeted for launch in the mid-2020s but remains in budget limbo as of the time of this writing, so it's unclear if it will ever get off the ground.

What does "nearby" mean? Well, that depends—on the brightness of the parent star, for example, and how far away from it the planet is. But 40 light-years is a pretty reasonable maximum-range estimate for JWST, Lisa Kaltenegger, director of Cornell's Carl Sagan Institute, said. We already know about a few possible abodes for life inside that bubble, including the roughly Earth-sized worlds Proxima b (4.2 light-years away), Ross 128 b (10.9 light-years), and the TRAPPIST-1 planets (39.6 light-years). And it's likely that we'll find more soon; our exoplanet cup continues to overflow.

Now, JWST won't be able to study all of these worlds; its atmospheric investigations will be limited to planets that "transit," or cross the face of their host star from the telescope's perspective. The TRAPPIST-1 planets transit, but Ross 128 b does not. As of the time of writing, the jury was still out on Proxima b.

Anyway, if JWST picked up methane in TRAPPIST-1f's skies, Kaltenegger and her colleagues would doubtless be excited, but they wouldn't exactly be doing cartwheels. Not all of Earth's methane comes from microbes and cow butts—there are geological processes that can make this stuff as well, such as the interaction between certain rocks and hot water. The detection of lots of oxygen, on the other hand, would definitely cause a few gasps, because that's a much stronger biosignature. Earth's atmosphere is 21 percent diatomic oxygen (O_2), and pretty much all of it comes from plants and photosynthetic microorganisms. If an evil genius—probably one with a name as glorious as Vesto Slipher—somehow wiped out all of these green guys tomorrow, the oxygen in Earth's air would likely disappear within a few million years, getting locked away in other molecules such as water and carbon dioxide.

But some researchers would still keep their feet on the ground, waiting for additional evidence to cover for exoplanet diversity. After all, weird chemistry on a weird planet may let nonbiological O_2 stick around for longer than we ever thought possible. For example, radiation from the parent star might split apart water or carbon dioxide molecules high up in the atmospheres of some worlds, generating large amounts of oxygen without the need for spinach or skunkweed. Kaltenegger (and many other scientists) would love for JWST to see oxygen swirling around with methane. Neither of these gases can last long in each other's presence—they react together easily, forming carbon dioxide and water—so spotting them both at the same time would be a big deal.

"From the knowledge that we have, if we find oxygen with a reducing gas on a planet that's in the habitable zone, we have no other explanation, except for it to be biology," Kaltenegger said. (Reducing agents lose electrons, and oxidizers gain them, in redox chemical reactions. In the methane-oxygen reaction outlined above, methane is the reducing agent and oxygen is the oxidizer.)

Of course, there's no guarantee that any aliens close enough to spy on will practice photosynthesis, or that, if they do, they'll churn out lots of oxygen in the process. After all, some weirdo photosynthetic bacteria here on Earth generate sulfur as a by-product instead. Maybe the only inhabitants of TRAPPIST-1f are rock-eating microbes that live 500 feet underground and the even tinier, dog-shaped microbes they keep as pets. Earth actually provides a pretty good cautionary tale in this respect. Photosynthesis seems to have evolved here pretty early on, but oxygen didn't start accumulating in our air until 2.4 billion years ago, and it may have been detectable from afar for only the last 600 million years or so—just 15 percent of the history of life on Earth. A better biosignature recipe for the early Earth would be methane and carbon dioxide, in the absence of carbon monoxide, some researchers have suggested.

We shouldn't get too locked in on oxygen, some scientists stress. To help alien hunters keep an open mind, a couple of years back MIT astrophysicists Sara Seager and her colleagues William Bains and Janusz Petkowski put together a list of 14,000 potential biosignature gases. If you scrape together a bit of information about an alien planet, you can winnow that imposing slab of substances down to a few promising targets. For example, ammonia (NH_3), chloromethane (CH_3Cl), and nitrous oxide (N_2O, the wondrous gas that makes whipped cream in a can possible) are good potential biosignatures on rocky worlds with hydrogen-dominated atmospheres, Seager and her colleagues have found.

Sorry if all of these possibilities and caveats are making your head spin a bit. But there's just so much to consider—so many different types of alien worlds, so many ways for their native creatures to make a living. Unfortunately, all of that diversity and uncertainty don't bode well for those of us who like our Earth-shaking, era-shaping discoveries to be as clean and sharp as an army haircut.

"We're never going to be 100 percent sure," Seager said. "We may be 99 percent sure, or 90 percent sure.

But it's not going to be perfect, never, no matter what we do, just looking at the atmosphere."

Bains agreed. And, he stressed, this percentage game is the norm for scientific findings, which are assessed according to how likely they are to result merely by chance. This is the famous p-value. If p is greater than 0.05, there's more than a 5 percent probability that the effect measured was due to chance alone, and a 0 percent probability that the study will get published in a decent journal.

"The really interesting thing is this: what if we find five planets with an 80 percent chance [of life]?" Bains said. "We're pretty sure now that there is life in the universe. So the headline announcement is, 'Scientists prove there's life in the universe.' And then someone like you"—meaning me—"comes along and says, 'Fantastic! Where?' And we say, 'We can't tell you.' That's going to be tricky."

Alien biospheres could betray their existence in other ways as well. In December 1990, NASA's Galileo spacecraft zoomed past Earth, getting a speed-boosting gravitational assist on its way to Jupiter. During the flyby, Galileo scrutinized our planet for signs of life, conducting what project leader Carl Sagan and his team called a "control experiment" for future ET hunts. The probe detected oxygen and methane as well as narrow-band radio signals, the latter solid evidence of intelligent life, unless they're coming from local sports-talk stations. And Galileo spotted something else—a pattern in the sunlight reflected by our planet. There was a big spike in bounced-back light just beyond red wavelengths, which are the longest ones that human eyes can see. (But we don't want to get too people-centric. Rattlesnakes and other pit vipers, as well as some boas and pythons, can see longer wavelengths—infrared light, or heat. This adaptation allows them to spot warm-blooded prey even in complete darkness.)

This phenomenon is called the red edge, and it's a result of photosynthesis. Plants and photosynthesizing microbes don't use infrared light, so these wavelengths get bounced back into space. Something similar could be happening on exoplanets with their own light-eating creatures, astrobiologists reason. And they're keeping an open mind: the edge on alien worlds could be green or blue or purple, depending on what pigments their plants use for photosynthesis.

There are even more exotic ways to parse alien light. Some corals here on Earth emit green or red light when exposed to ultraviolet radiation, presumably to help prevent sunburn. Rather than absorbing potentially damaging UV light into their tissues, the corals are shifting it to safer, lower-energy wavelengths, the thinking goes. If some alien creatures do the same thing, it may be possible to spot a gorgeous glow washing over their worlds in the wake of a powerful solar storm, Kaltenegger and her Cornell colleague Jack O'Malley speculated in a recent paper.

We're still pretty new at this whole life-hunting game. We don't yet know for sure which tools, search strategies, target planets, or biosignatures should occupy most of our time and attention.

"We stumble forward," said Thomas Zurbuchen, head of NASA's Science Mission Directorate. "It's like we're getting into a dark room. We're lighting up corners of it, but every once in a while, there are entirely new rooms attached to that room. And I think that's going to be what's going to happen in the future, too."

BITS AND PIECES

The ET hunt is also zeroing in on potential bits and pieces of the putative creatures themselves—a more direct approach that's especially important for worlds whose inhabitants wouldn't leave any sort of imprint in the air, like the buried-ocean moons Europa and Enceladus.

For example, both NASA's Mars 2020 rover and the European-Russian ExoMars rover will search for chemofossils—complex organics that may once have been part of ancient microbes—when they touch down on the Red Planet in 2021. Perhaps one of these robots will stumble across a pile of amino acids or a clump of the fatty molecules that cells use to build their membranes.

By the way, not just any old amino acids would do, since these protein building blocks occur naturally throughout the solar system. Astrobiologists would also want to see evidence of "handedness" in the sample. Amino acids and many other biomolecules come in two mirror-image forms, which line up just like your hands do. Here on Earth, life uses only left-handed amino acids (and right-handed sugars), so finding a similar imbalance on Mars would be a strong indication of alien life.

Then there's DNA. Two different research groups— one led by synthetic-life pioneer Craig Venter and the other (which includes Gary Ruvkun) based at MIT—are developing gene sequencers for potential use on Mars and other alien worlds. DNA is not a good chemofossil candidate; it doesn't preserve well at all. So, if one of those little machines eventually flew to Mars and got a hit, it would be a sign of very recent and probably even present-day life. The shared use of DNA would strongly suggest that it's related to us—a not entirely outlandish prospect, as we saw in chapter 2.

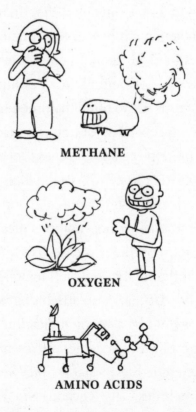

METHANE

OXYGEN

AMINO ACIDS

What about weird life—creatures that use a whole different set of biomolecules than Earth creatures do? Well, we may still be able to flag whatever their DNA equivalent is, as long as those aliens rely on liquid water as a solvent. Biochemist Steven Benner and some like-minded folks have argued that all genetic molecules capable of supporting Darwinian evolution—a key part of NASA's life definition, remember—will be like DNA in one key respect: they'll have a repeating electric charge going up and down their long backbones. Such a charge keeps the molecule from folding in on itself like a wadded-up tissue, among other things. DNA's charge is negative, but the one on the alien molecule could be positive; it doesn't matter. It should be possible to design an instrument that can detect such extravagantly charged molecules, and then fly it to Europa, Enceladus, or other ocean worlds that may have hosted a second genesis of life very different than that of Earth, Benner and his colleagues have said.

Unfortunately, we have to inject some pretty big caveats into this part of the discussion. Looking for biomolecules, preserved or fresh, is tricky work for a robot on a cold and distant world. There will probably be some ambiguities in any positive detection, and other researchers will certainly find them. Scientists are a skeptical lot in the most ordinary of times—that mindset is part of the job description—and they'll pick over any putative discovery of alien life with the zeal of hyenas at a giraffe carcass, given the stakes of the claim.

We don't have to speculate about this. The ET-vetting apparatus has kicked into gear on two high-profile occasions—after the Vikings' work in the mid-1970s and in 1996, when a team of researchers reported finding evidence of life in the Mars rock ALH 84001 (see chapter 10).

"I think what you really need in order to have proof is to go with a microscope and see bacteria wiggling to and fro and waving back to you," astrobiologist Dirk Schulze-Makuch said. "And even then, it would be difficult. People would still say, 'Oh, maybe that was contamination. Maybe you brought all those bacteria from Earth.'"

Astrobiologists are therefore keen to haul newly collected, pristine samples from alien worlds to Earth, so they can look for life with all the fancy equipment available in their labs. That wish might come true, but it's hard to say when: the Mars 2020 rover will collect and cache promising samples, but, as of this writing, there is no mission on NASA's books to grab them and bring them to Earth.

SIGNS OF INTELLIGENCE

We've ventured pretty far from the Lowell-lands, but it's now time to head back there, because the hunt for intelligent, if not necessarily thirsty, aliens is still very much on. And it's come a long way since canals topped the list of primo biosignatures.

When you see the acronym *SETI*, you probably think "radio signal." There's a good reason for that: though some projects try to spot super-bright laser flashes, most of the SETI efforts over the past six decades have used radio telescopes, like the big dishes at Green Bank in West Virginia and the Arecibo Observatory in Puerto Rico, to look for narrow-band emissions. These are signals that cover a very small frequency range, such as the ones you can pick up on your AM or FM dial.

"A lot of energy at one frequency—that is not something that nature does," said Dan Werthimer, chief scientist at the University of California, Berkeley's SETI Research Center. Signals generated by clouds of interstellar gas, supermassive black holes, and other naturally occurring objects are much more dispersed, akin not to an execrable '70s-rock station but to the oasis of blaring static just past it on the dial.

Such a narrow-band signal could be meant for us, or it could be an alien-to-alien transmission that we just happen to intercept—a Grey commander dressing down a subordinate for failing to meet his monthly probing quota, say. Whatever the case, if astronomers spotted one and confirmed that it came from deep space, there would be cartwheels down carpeted halls and probably a trip to the fancy-box shop to find a velvet-lined home for that Nobel Prize medal. But the confirmation part is key, because false alarms are common; Earth and its environs are very radio-noisy. For example, in May 2015, Russian astronomers picked up an intriguing signal that initially seemed to be coming from a star about 94 light-years away. By the next August, however, they had determined that the source was terrestrial—probably a Russian military satellite.

The chances of a detection are probably better now than they they've ever been, because observing gear and especially data-analysis technology have improved so much. Back in SETI's early days, Werthimer said, astronomers could scan just 100 or so individual radio channels simultaneously. Now they can do about 100 billion. And the chronically underfunded enterprise recently got a big infusion of cash, courtesy of Russian-born billionaire Yuri Milner. Milner is funding Breakthrough Listen, a $100 million, 10-year-long project that's searching for radio signals from the center of the Milky Way and the galactic plane, 100 of our nearest neighbor galaxies, and the 1 million stars closest to Earth. Breakthrough Listen team members are also doing some "optical SETI" work, looking for powerful laser flashes. (The Breakthrough Listen science program is based at the Berkeley SETI Research Center; Werthimer is a team member.)

But that doesn't mean the chances are *good*. SETI scopes generally look at only a tiny fraction of the sky at a time, either during targeted searches of individual stars or plodding all-sky surveys that can take months to complete. This strategy works well for detecting signals that are constantly blaring—the scopes will get to them eventually—but it's not so great for rare events. If ET is sending out a sort of lighthouse beam that sweeps over Earth just once a year, for example, we probably won't spot it. For this reason, SETI astronomers have long dreamed of an "all-sky, all the time" system. And some people are now working to make it happen, at least for light in the optical range. Werthimer is part of a project called PANO-SETI, which aims to build geodesic domes at multiple locations around the world. The roof of each dome would be covered by 126 special hexagonal lenses, which together would collect light from huge swaths of sky.

There are lots of different frequencies to search, from long and lazy radio waves all the way up to the gamma range, radiation so energetic it'd sizzle up your guts like a porterhouse.

There's another big issue with SETI searches, whatever wavelength of light they may target: we have no idea if we're looking for the right signals. We're hunting for radio and laser pings because those are technologies that we ourselves use and understand; there's no guarantee that ET is on the same page. In fact, as noted in chapter 1, there are very good reasons to think that intelligent aliens will be much more advanced than we are—perhaps so advanced that we cannot conceive of, let alone detect, the signals they're sending out through the galaxy.

SETI scientists have thought about this, of course.

"I think it's very hard to predict what a civilization might do," Werthimer said. "It wouldn't surprise me if what I'm doing and other people are doing in SETI is completely wrong, and in 100 or 200 years people will laugh—just the way that we laugh now at the early SETI ideas, like building fires and using mirrors."

But the uncertainty and the prospect of being mocked by future generations[1] shouldn't keep us from trying. Werthimer added: "You've gotta do what you know how

---

1   Who cares? They're going to laugh at our clothes and hair anyway.

to do, and use the physics and the science and the technology of the day."

Ah, technology. That's really what SETI scientists are after—not signs of intelligence, per se, but evidence of alien technology. Indeed, SETI astronomer Jill Tarter has called for the field to be rebranded as "the search for technosignatures," because *intelligence* is such a slippery and loaded term. We have a hard enough time defining it here on Earth—Are dogs intelligent? Are rats?—so how do we go about searching for it out in the cosmos?

And there are many technosignatures besides radio or laser signals that we could theoretically pick up. For example, if one of those big future telescopes detected a complex industrial chemical—say, chlorofluorocarbons, the hairspray stuff that famously ate a hole through Earth's ozone layer—in TRAPPIST-1f's air, we could be fairly confident that some "advanced" creatures lived there. And if all the TRAPPIST-1 worlds had the same pleasant temperature, despite their different distances from the star, many astronomers would infer some large-scale climate engineering. And worms or slugs would probably not be responsible.

There are some other, more famous possibilities. For example, in 1960, theoretical physicist Freeman Dyson suggested looking for evidence of huge solar-panel arrays around stars. Super-advanced alien civilizations might build such Dyson spheres, as they came to be known, when their power needs eclipsed those available on the surfaces of their home worlds, he reasoned. And in 2015, astronomers announced something that fit the rough description. A star about 1,500 light-years from Earth called KIC 8462852—better known as Tabby's star or Boyajian's star, after the leader of that study, Tabetha Boyajian—had dimmed dramatically several times over the previous half-decade or so, one time by 22 percent. Researchers didn't assume that an orbiting, half-built Dyson sphere was blocking all that light; they offered up a number of natural explanations, all of which were viewed as more likely. But some did mention an "alien megastructure" as a viable possibility, which is pretty exciting in itself.

Alas, the megastructure theory is pretty much dead now. Further observations of the weird star's light strongly suggest that the culprit is a cosmic dust cloud.

Similarly, some astronomers entertained a rather outré idea in the fall of 2017 after a telescope in Hawaii spotted, for the first time ever, an object from interstellar space zooming through our solar system. This body, later named 'Oumuamua (Hawaiian for "scout"), is distinctly odd: it's about 800 feet long by 100 feet wide and looks like a giant, rocky needle.

'Oumuamua is so odd, in fact, that it might be an alien spaceship of some sort. And don't just take my word for it.

"It may be that there are these probes that we haven't noticed until recently, and once we develop good enough technology—like, for example, the LSST—we'll suddenly see all of these things," said Avi Loeb, the chair of Harvard University's astronomy department, referring to the Large Synoptic Survey Telescope. This powerful optical instrument is being built in the Chilean Andes; after it comes online in 2019 or so, the LSST will survey the entire sky every few nights.

"So I think it's possible that, once we reach a threshold and we look for the right signals, we will figure out that they're not dumb," Loeb added. "They are not just sending radio signals; they have much more sophisticated technology."

Loeb isn't making a claim here; he's just saying it's possible. And a number of astronomers agree. Several different teams—including folks based at the SETI Institute and Breakthrough Listen—have turned their radio dishes toward 'Oumuamua, hoping for a ping. They've gotten nothing so far; the big, mysterious needle rock is zooming silently toward the solar system's outer reaches, a masked thief disappearing into the darkest of nights.

'Oumuamua and Tabby's star are anomalies—things that just don't look right, like an unattended bag on the subway or a Shar-Pei. Astronomers should be on the lookout for more such weird stuff, Werthimer said, and that hunt should extend into the past as well. He said that it's possible that some telescope on Earth or in space has already spotted evidence of ET, but the prize nugget is buried under a mountain of other data. Data-mining algorithms could well dig it up.

Given that we really don't know what we're looking for, it may not be a SETI scientist who ends up making the longed-for discovery. For example, evidence of advanced alien life may come from a physicist studying dark energy who notices an odd glitch in his or her data.

"If you look at the long-term history of big discoveries in astronomy, about half of them are made completely by accident," Werthimer said.

He cited pulsars, the fast-spinning, super-dense stellar corpses that blast beams of high-energy radiation out into the cosmos. They emit these beams constantly, but the spinning makes them seem to pulse. Think of a lighthouse sweeping its rays around a rocky coast; the analogy works, and lighthouses are picturesque and evoke childhood vacations to seaside towns, when life was simpler and better.

A Cambridge University PhD student named Jocelyn Bell discovered pulsars in 1967, after analyzing the results flowing from a newly built radio telescope. She noticed a weird signal that kept repeating every 1.33 seconds. It initially stumped Bell and her doctoral adviser, Antony Hewish, and they dubbed the mysterious source Little Green Man 1 (LGM-1), because it was consistent with the sort of signal you'd expect from intelligent aliens. Bell, no SETI scientist, might have been the person to finally find ET.

But then she spotted another such flickering ping coming from another spot in the sky, and she figured that ET wasn't responsible—instead, it was some natural cosmic phenomenon that nobody had seen before. (The find, by the way, earned Hewish and fellow astronomer Martin Ryle the 1974 Nobel Prize in Physics. Bell was shut out, a decision that some of her colleagues found unjust but she has said she accepts, given her lowly status as a research student at the time.)

Astrophysicists are grappling with something of a pulsar/LGM-1 situation right now, as a matter of fact— something called fast radio bursts. These are quick, short blasts of radio light coming from deep, deep space, some of which repeat. The first one was spotted in 2007, and astronomers have now racked up about two dozen of them. They may be coming from merging black holes or super-magnetic star corpses. It could be ET. Probably not, but who knows? It's possible.

Chapter 6

Where Is ET Hiding?

Sir William Herschel was the most famous astronomer of the late 1700s and early 1800s. The
German-born Brit made his bones in 1781 by discovering Uranus, the first new addition to our solar system's
planetary panoply since the days of togas and centaurs.
He followed that up by cataloguing thousands of deep-
space objects, figuring out that Mars's polar ice caps
change with the seasons, spotting Enceladus and several
other giant-planet satellites, discovering infrared radiation,[1] and making the case that there's life on the sun.

---

1   In 1800, Herschel measured the temperatures of different wavelengths
     of sunlight, which he had split using a prism. He determined that the in-
     visible light beyond the red end of the spectrum was even warmer than
     the wavelengths we can see. The European Space Agency named its

Yep, the sun. Like many other scientists of his day, Herschel thought creatures abounded on pretty much every decent-sized planet or moon. The great German mathematician Carl Gauss, for example, mused about using mirrors to signal our presence to the "lunarians" living on the nearest rock to Earth. "This would be a discovery even greater than that of America, if we could get in touch with our neighbors on the moon," Gauss wrote in an 1822 letter to the German doctor and amateur astronomer Heinrich Olbers, according to Michael Crowe's 1986 book *The Extraterrestrial Life Debate, 1750–1900*.

What does this have to do with the sun? Well, Herschel thought Earth's star was actually a planet. Solar radiation, he argued, beamed forth from a luminous shell surrounding the sun, beneath which floated a super-insulating layer of "planetary clouds." This belief stemmed from Herschel's interpretation of sunspots as fleeting glimpses of the clement solar surface, through gaps in the outer shell and cloud layers. (We know today that sunspots are temporary magnetically active patches cooler than the rest of the solar surface. But "cooler"

Herschel Space Observatory, which viewed the cosmos in infrared light from 2000 to 2013, after William and his sister Caroline, who collaborated with him frequently and made important discoveries of her own.

is all relative: sunspot temperatures are about 8,000°F, compared to 10,000°F. By the way, our star's temperature peaks in its outer atmosphere, or corona, at several million singeing degrees.)

The solar surface was therefore a nice place for "beings whose organs are adapted to the peculiar circumstances of that vast globe" to run and frolic and fight and fuss, Herschel wrote in a 1795 paper.

This sun stance was not shared by most of his peers, who figured that stars are different than planets in some pretty basic and important ways. (Spoiler alert: they were right.) The last two centuries of astronomy and heliophysics haven't exactly made us optimistic about the discovery of squat, hot-tempered solarians. But many other cosmic locales, both in our solar system and beyond, may indeed be abodes for life today. In this chapter, we'll briefly tour some of the most promising ones. To soothe the tormented ghost of Percival Lowell, we'll start with Mars.

RED BUT NOT DEAD?

Ancient Mars seems to have been a pretty decent cradle for life, as we saw in chapter 2. And many astrobiologists think that if microbes did indeed take root on the Red

Planet during its long-gone salad days, they're probably still scratching out a living there today.

Think about how tough Earth microbes are. Many aptly named extremophile species thrive at volcanic deep-sea vents, enduring temperatures as high as 250°F and pressures that would pop your head like a grape. Others happily while away the hours in hot and muddy little ponds with a flesh-melting pH lower than that of battery acid. And you know all about radiation resistance from *D. radiodurans*—"Conan the Bacterium," remember?

"Evolution might accept a less grandiose lifestyle than some kind of boreal forest, but [life] doesn't typically go away, especially if there's something warm and wet in front of it," said SETI Institute scientist John Rummel, who served as NASA's planetary protection officer from 1990 to 1993 and again from 1998 to 2006. "So, given that underneath Mars there's very likely to be warm and wet spots, I'm going for life on Mars."

That's right: though the Red Planet's dry and frigid surface is generally regarded as an even worse place to live than Las Vegas, there may be pockets of life down below—perhaps miles below, in and around deeply buried aquifers. It's possible that we've even caught a whiff of this life already, in the form of those methane plumes we talked about in chapter 5.

If these pockets do indeed exist, we may be able to study them up close someday, provided their water links up with cave systems that future human explorers or nimble spelunkobots can access.

Oh, and in case Rummel's former job title caught your eye: planetary protection isn't about lobbing grenades into swarming masses of alien invaders. Rather, it describes the effort to minimize the odds of contaminating other worlds with Earth life, or bringing ET microbes back here by accident. And this is a real concern, considering how hardy some microbes are; Conan the Bacterium seems perfectly capable of surviving the long, deep-space trek to Mars.

NASA therefore mandates that any landers or rovers set to explore "special regions"—potentially warm and wet spots where Earth bugs might be able to survive today—be sterilized to an exacting (and expensive) degree before liftoff. To date, only the Viking landers have been scrubbed so thoroughly. For example, under current rules, Curiosity isn't allowed to inspect the seasonal dark streaks on warm Martian slopes that some scientists think are caused by liquid water. This frustrates some astrobiologists, who'd like to relax the rules a bit. We should chase after Red Planet life in the places it's most likely to exist, using the tools we currently have at our disposal, they say. But others, like Rummel, are fans of a strict protocol, both out of respect for potential native alien ecosystems and because it could make future astrobiology discoveries on Mars and other worlds easier to interpret.

"The bottom line is, the search for alien life is too important, and too costly, to do a half-assed job," Rummel said.

VENUS

It seems absurd, almost Herschelianly so, to speculate about life on present-day Venus. This is the planet, after all, with surface temperatures around 860°F—hot enough to melt zinc or cadmium—and a carbon dioxide atmosphere 90 times thicker than the air of Earth. No lander has been able to survive this crushing hellscape for more than 127 minutes, so how could life even get a foothold, let alone make do for billions of years? (In case you were wondering, that record was set by the Soviet Union's Venera 13 probe back in 1981.)

Like Mars, Venus was very different long ago. Observations by a variety of spacecraft suggest that the second planet from the sun was a balmy ocean world early in its history, as habitable as the Turks and Caicos. But Venus's tight orbit eventually caught up with it.

Sun-like stars are like people—they're relatively dim when young but brighten as they mature. So, after our sun passed through its awkward teenage years, Venus warmed and its oceans began boiling away into the atmosphere. Water vapor is a potent greenhouse gas, so the planet then heated up even more. That caused more of the ocean to evaporate, resulting in an even hotter Venus, and so on. Venus ended up losing virtually all of its water to space—via stripping by the solar wind, as happened at Mars as well—and carbon dioxide built up in the air, eventually coming to dominate the atmosphere. (It's this heat-trapping CO_2 that keeps Venus so toasty today.)

Also like Mars, Venus may still boast a potentially habitable environment—but this one lies above the ground, not beneath it. About 30 miles up in the Venusian skies, temperatures and pressures are pretty Earthlike. Sure, Venus clouds are made of sulfuric acid, but microbes are tough, remember? So, a number of researchers, Carl Sagan among them, have raised the possibility that life could exist up there. Astrobiologist Dirk Schulze-Makuch has even suggested that the dark streaks visible in UV images of Venus's atmosphere may be caused by microbes cocooned in a self-secreted sulfurous sunscreen.

Schulze-Makuch thinks Venus probably hosted life

billions of years ago in the form of native-born microbes and/or beasties that journeyed there on rocks from Earth. (Venus is closer to the sun than Earth is, so many more of our rocks have made it to Venus over the eons than the other way around.)

Given life's tenacity—"It's like a cockroach in your house," Schulze-Makuch said—Venusians may survive still, if their long-ago ancestors had time to move from the ocean to the air.

But that's the rub: nobody knows exactly when Venus turned into the Planet of Eternal Suffering, or how long this transformation took. "If it was one catastrophic burst, I can imagine that all life was wiped out, everything was sterilized, and that was it," Schulze-Makuch said. "But if it didn't happen in that catastrophic scenario, then I find it likely that life would still be hanging in."

We could theoretically find such floating life, he and others have suggested, by launching a sample-snagging balloon mission to Venus's skies (and, ideally, bringing those samples back to Earth). Don't laugh: Venus balloons are an actual thing. The Soviet Union's Vega 1 and Vega 2 missions used balloon probes to gather lots of data about the Venusian atmosphere back in 1985. Neither mission looked for life, though.

EARTH

What's Earth doing on this list? Well, we could all be Martians, as discussed in chapter 2. And more than 12 million Americans believe that (presumably alien) lizard people run the country, according to a 2013 poll. But those aren't the only reasons. Some scientists have argued that a completely separate tree of life could lurk undetected on our planet, in a "shadow biosphere" right under our noses. (Semantics note: if shadow biospherians exist and evolved here on Earth, they would be aliens of a sort, but not extraterrestrials.)

Like all the crazy ideas in this book, this isn't such a crazy idea when you think about it. Recall that "normal" microbes—the kind whose descendants include leeches and lizards and leopards and us—got going very quickly on Earth, suggesting that the whole origin-of-life thing may not be such a big deal. So maybe it happened twice here, or three times, or ten.

If second-genesis creatures exist, they may go about life's business in a fundamentally different way—using a molecule other than DNA or RNA as genetic material, for example, or a different set of amino acids to build proteins. Indeed, such key biochemical differences could explain why we've found no evidence of "weird life" to date: standard microbe-studying methods are geared toward detecting and culturing life as we know it.

So the hunt for the shadow biosphere centers on weirdos—microbes that do odd things or live in especially harsh or out-of-the-way places, like California's super-salty, arsenic-rich Mono Lake. A decade back, a team of researchers scoured Mono Lake for signs of weird life and ferreted out an extremophile bacterium that seemed to incorporate arsenic into its DNA instead of the normal phosphorus. *Seemed* is the operative word here, however. Other research teams later determined that this tough little bug, dubbed GFAJ-1, does indeed need phosphorus; it's just very arsenic-resistant.

Weird life might not be restricted to such exotic locales. In fact, you may even have seen evidence of it yourself, especially if you like hiking through canyonlands or the Karoo in South Africa. Rocks in arid environments around the world are commonly coated with a dark, sometimes shiny layer known as desert varnish. It's perhaps most obvious at petroglyph sites, where ancient artists often scratched their representations of deer and dancers through the stuff.

Surprisingly, nobody knows for sure what desert varnish actually is. Analyses have revealed that this ultrathin patina contains clay, manganese, and iron, along with organic molecules, but its origin remains a mystery. Some scientists think the stuff is made of deposits left over the eons by manganese-munching microbes. The elusiveness of these putative bugs has led to speculation that they may be weirdos, brothers from a different evolutionary mother.

ICY OCEAN WORLDS

If you think that the drying of Venus and Mars left Earth as the solar system's only ocean world, find a buzzer and press it, because you are wrong.

There's plenty of liquid water sloshing around in the outer solar system; it just happens to be buried under miles of ice. Scientists are pretty sure that subsurface oceans exist on the ice-covered Jupiter moons Europa,

Callisto, and Ganymede, as well as the Saturn satellites Enceladus and Titan. And some observations suggest they may also exist on two additional Saturn moons, Mimas and Dione, as well as Neptune's big satellite Triton and the dwarf planet Pluto. (Icy, battered Mimas has a huge crater that makes it look like the Death Star, or a giant, staring space eye. Look it up—it's really cool.)

And calling them oceans is no exaggeration; Europa's is thought to contain twice as much water as all of Earth's seas combined.

You may be wondering why such seas don't freeze solid, given how cold it is on these dark and distant worlds. Don't buzz yourself—that's a good question! The answer lies in a phenomenon called tidal heating. Basically, the powerful gravity of these moons' parent planets stretches and squishes their insides, generating loads of heat via friction. This happens because the moons' orbits aren't perfectly circular, so the tidal bulge caused by this gravitational tugging changes over time. Tidal heating is also responsible for churning the insides of Jupiter's sulfur-spewing moon Io, the most volcanic object in the solar system. (Pluto has no parent planet, of course; the tidal forces it experiences are generated by its five moons.) Heat given off by the natural decay of radioactive elements deep in the ocean worlds' cores may also play a role, especially for Pluto and Triton.

Two of these alien oceans get the brunt of the astrobiological attention—those of Enceladus and Europa. That's partly because these two seas are thought to be in contact with the rocky cores of their respective moons, allowing a variety of complex and possibly life-sparking chemical reactions to take place—and partly because Enceladus and Europa are just really, really cool. (The oceans of Callisto and Ganymede are likely sandwiched between layers of water ice. You don't have to be a chemist to intuit that water + water ice = boring. The interior structures of the other buried-ocean worlds generally aren't well understood, so it's tougher to speculate about them.)

Let's establish the bona fides of Enceladus, the most reflective world in the solar system. In 2005, NASA's Saturn-orbiting Cassini spacecraft discovered geysers of water vapor and other stuff blasting at 800 miles per hour from deep, snaking "tiger stripe" fissures near Enceladus's south pole. There are more than 100 of these jets, and their combined output forms a plume that wafts hundreds of miles out into space—so far out, in fact, that its material makes up Saturn's E ring. The geysers are also just a joy to behold. In some long-range Cassini photos, their bright, icy tendrils make Enceladus look like a spherical rocket blasting off, or a bulbous-belled young jellyfish bobbing about in a nighttime sea.

Over the course of its 13 years circling Saturn (which came to an end with an intentional death plunge into the ringed planet in September 2017), Cassini flew through the plume many times, tasting Enceladus's ocean—because that's where the geyser material is coming from. The probe wasn't outfitted with any life-detection gear, but it did spot some stuff that got astrobiologists very excited. Simple organic chemicals, for example, as well as molecular hydrogen (H_2) and a surprising amount of methane (CH_4). These latter two substances hint at the presence of hydrothermal vents on Enceladus's seafloor, the same energy-rich environments where life may have gotten its start here on Earth. Indeed, the hydrogen itself could help sustain life: many Earth microbes greedily gobble it up. These creatures are called methanogens, because their metabolism revolves around reacting molecular hydrogen with carbon dioxide to produce methane (and energy). All methanogens are archaea—microbes similar to, but evolutionarily distinct from, bacteria. Many extremophiles are archaea as well.

All this, and Enceladus is just 313 miles in diameter— about the same as the distance across Arizona.

Before you get too excited about Enceladus life, let me throw a little cold buried-ocean water on the prospect. Some recent research suggests that the satellite, some of

Saturn's other inner moons, and its iconic ring system may be quite young. That is, they might not have formed along with Saturn 4.5 billion years ago, but rather coalesced 100 million years ago or so from the shattered remnants of predecessor moons destroyed by giant impacts. This hypothesis is based on peculiarities of the orbits of Enceladus and other inner moons (not Titan, which is widely regarded as ancient), as well as the surprising brightness and low mass of Saturn's rings. It makes a certain amount of sense: why else would Saturn's rings be so bright and dramatic, compared to the drab and shamefully degraded ring systems of Jupiter, Uranus, and Neptune?

If this idea is on the money, then life hasn't had all that much time to take root on Enceladus. Somewhat paradoxically, the large amount of H_2 that Cassini spotted in the plume may actually suggest that life doesn't exist on the shiny moon, or at least that it's not there in profusion. Here on Earth, molecular hydrogen gets gobbled up quickly by methanogens, so you presumably wouldn't see much of it in an Earth plume (if we had one).

Now on to Europa. The Jovian satellite is much bigger than Enceladus—1,940 miles in diameter, nearly as large as Earth's moon—and therefore has a lot more water. Scientists think the global Europan ocean may be a whopping 60 miles deep. (The deepest known point on Earth's seafloor, by comparison, is a measly 6.8 miles.) It's unclear just how close this sea gets to the surface; estimates of the ice shell's thickness range widely, from 1 mile or less to about 20 miles. (Nobody knows how thick Enceladus's ice shell is, either. But it's probably much thinner than Europa's.)

Europa looks a bit like a bloodshot eye whose iris and pupil have been scooped out with a spoon, or at least pounded into a gelatinous smear (with the back of the spoon, I guess). The capillaries in this analogy are long cracks in the ice that are filled with "brown gunk"—the actual technical term scientists use for the stuff. Lab experiments suggest the gunk is salt from the buried ocean, which has made it to the surface and been discolored by the intense radiation environment there. And intense it is: Europa, like its fellow travelers Io, Callisto, and Ganymede, orbits within Jupiter's radiation belts and is therefore constantly bombarded by fast-moving charged particles.

Thus Europa's surface wouldn't be a safe place to visit; you may as well take a vacation inside the X-ray scanner at the airport. But this powerful radiation could actually make the moon's buried ocean a more livable place. It splits apart molecules of water ice, freeing up superreactive oxygen, which little Europan beasties swimming around in the dark, cold depths could use as a chemical energy source.

How could this oxygen get all the way down to the ocean? Another good question! (You can throw away that buzzer now.) Well, Europa seems to have plate tectonics, like Earth—but the moon's sliding, scraping slabs are made of ice, not rock. Deep-diving slabs could be dinner plates, perhaps even delivering enough chemical grub to support creatures big enough to see with the naked eye (if you had a powerful waterproof flashlight). And these diving plates wouldn't necessarily have to make it all the way down to the ocean, either. Scientists think there are probably small pools of liquid water scattered throughout Europa's ice shell; these could be habitable environments, too.

Europa may also have geysers—though, if so, they're probably intermittent rather than constantly blasting like the jets of Enceladus. The Hubble Space Telescope has spotted an apparent water vapor plume wafting away from Europa on several occasions, a find that astronomers were still trying to nail down as of the time of this writing.

TITAN, TOO

Titan also seems to have a buried ocean, but most speculation about life on the 3,200-mile-diameter Saturn moon focuses on its surface. That's partly because Titan's interior is poorly understood, but mostly because what's going on up top is so interesting.

As mentioned in chapter 3, the frigid moon is dotted with seas of liquid ethane and methane, some of which are even bigger than Lake Superior (but far calmer; Cassini's measurements suggest the exotic waves are generally less than 1 inch high). Indeed, Titan's surface has hundreds of times more hydrocarbons by volume than all of Earth's oil and natural-gas reserves combined. The stuff literally falls from the sky; it's the basis of Titan's weather system.[2]

The big moon has a thick, nitrogen-dominated atmosphere, like Earth. You couldn't breathe it, at least not happily; there's no oxygen to speak of. But there are a lot of organic molecules swirling around in that thick, smoggy air and fluttering down onto the frigid surface below. There's lots of chemical energy available, too—molecular hydrogen, just like in Enceladus's ocean.

---

2   Thankfully, Titan oil rigs won't be economically feasible anytime soon.

But even if our putative Enceladus and Titan microbes ate the same food, they'd be very different beasties. Enceladus (and Europan) organisms would probably be more or less recognizable to us—carbon-based critters that use water as a solvent. Anything swimming in Titan's hydrocarbon seas would be weird life, and it's therefore hard to know what it would look like.

"The one thing we could say about Titan life with certainty is that it uses energy. That's about it," astrobiologist Chris McKay said. "Once we move away from water as the medium for the biochemistry, we are completely ungrounded in terms of looking for genetics, or structures, or whatever."

This reasoning doesn't necessarily apply, of course, to any microbes that may make a living in Titan's suspected subsurface ocean of liquid water.

GETTING SOME ANSWERS

Scientists are already putting some of the above speculation to the test. The Trace Gas Orbiter has started sniffing for Mars methane, as mentioned in chapter 5. And Europa is about to get a close-up: both NASA and the European Space Agency plan to launch missions to the ocean moon in the 2020s.

NASA's probe, called the Europa Clipper, will orbit Jupiter and characterize the moon's buried ocean during dozens of close flybys using a variety of instruments, in an attempt to determine just how habitable it really is. The Clipper will also scout out promising touchdown spots for a life-hunting Europa lander, a future project that Congress has ordered NASA to develop. The European entry, called JUICE, focuses on giant Ganymede but will also study Callisto and Europa.

JUICE is short for Jupiter Icy Moons Explorer, a mild case of lexical gerrymandering. There are far more tortured space-mission backronyms. My vote for the most egregious is NASA's asteroid-sampling mission OSIRIS-REx (Origins Spectral Interpretation Resource Identification Security-Regolith Explorer). No animus toward the mission itself, though: OSIRIS-REx is really cool. If all goes according to plan, it will return a sample of dirt and rock from the potentially dangerous asteroid Bennu to Earth in September 2023.

Europa Clipper and JUICE are on the books. A dedicated Titan mission may soon join them: a project called Dragonfly is one of two missions, along with a comet sample-return spacecraft called CAESAR (Comet Astrobiology Exploration Sample Return—a pretty respectable acronym, actually) under consideration for launch in the 2025 timeframe. If you like awesome stuff, you'll love

Dragonfly. It's a mini helicopter that would land on Titan and then cruise around to various spots, studying the moon and its life-hosting potential. One of Dragonfly's instruments would allow it to sniff out differences in the concentration of hydrogen gas—a possible signature of life on the huge and smoggy moon. We'll know soon if Dragonfly will get off the ground: NASA is scheduled to announce its pick in late 2018. (Dragonfly wouldn't be the first probe to touch down on Titan; Europe's Huygens probe, which traveled to the Saturn system with Cassini, landed on the moon in January 2005. But Huygens didn't carry any astrobiology instruments.)

Then there's Enceladus. McKay leads a team that's developing a possible plume-sampling mission called Enceladus Life Signatures and Habitability, which was a contender for the 2025 launch slot. ELSAH didn't make the finalist cut, but NASA continues to support it with some technology development funds.

Let them all fly, I say! Seriously, if you're a billionaire who wants to make a mark and leave a legacy, follow the lead of Yuri Milner and Paul Allen (whose donations allowed the SETI Institute's Allen Telescope Array to be built) and fund the search for alien life. If your project ends up being the one that finds something, people will remember and laud you forever, no matter how many autoworkers and arthritic Walmart greeters you may have put out of work with your mergers and acquisitions along the way. There could be a statue of you at the Smithsonian, or at least at the Roswell UFO Museum. Think about it!

FARTHER OUT

We've already moved a fair bit in this brief tour, from the sun to Venus and Mars, and then out to Jupiter and Saturn. Let's keep going—far beyond our own provincial edge, to alien worlds orbiting alien suns.

As we learned in chapter 1, there are billions of potentially life-supporting planets out there in the Milky Way galaxy alone. A fair number orbit sun-like stars. And these are the gold standard for habitable planets; the only life we know of took root on a world that orbits such a star, after all.

But we may be an outlier. The vast majority of habitable real estate in the Milky Way can be found around red dwarfs, the small, dim stars that make up about three-quarters of the galaxy's stellar population. Proxima b, Ross 128 b, the TRAPPIST-1 worlds—all of them orbit dwarf stars. This is not a trivial distinction, because just how habitable such worlds may be is a subject of intense debate.

For example, red dwarfs are incredibly active in their

youth, blasting out powerful flares with frightening frequency. Some studies suggest that these outbursts likely strip away the atmospheres of planets that would otherwise have a shot at supporting life, effectively sterilizing them. Then there's the proximity issue. Because red dwarfs are so dim—TRAPPIST-1 is just 8 percent as massive as the sun—their habitable zones lie much closer in than those of sun-like stars. As a result, most orbiting planets warm enough to host life as we know it are "tidally locked," always showing the same face to their host star, just as the moon always shows one face to Earth. (It takes the moon the same amount of time to complete one orbit of Earth as it does to complete one rotation along its axis—about 27 days. So the far side is forever lost to those of us confined to the planet.) And that may be a pretty big deal: the daylight side is likely scorching hot, while the nighttime side is consigned to an eternity of Antarctic chill.

Or maybe not. Some modeling studies suggest that a thick enough atmosphere could transport heat around tidally locked worlds, making at least some parts of them capable of supporting life as we know it.

In short, we really don't know how habitable red dwarf planets may be; the exoplanet revolution is still in its early days, after all. Astrobiologists are keen to get to the bottom of this issue, because red dwarfs are incredibly long-lived: they keep shining for trillions of years, meaning that life theoretically has loads and loads of time to take root and evolve on the worlds orbiting them. (Sun-like stars, by contrast, live for just 10 to 12 billion years.)

But, as far as hunting for life goes, star type is a secondary consideration. Far more important are the exoplanet's proximity to Earth and whether or not it transits its host star from our perspective—two characteristics that dictate how easy a world is to study. That's how Lisa Kaltenegger, director of Cornell's Carl Sagan Institute, sees it, anyway.

"So, my favorites list is: the closest planet ever, Proxima b. And then, I think, Ross 128 b is the second one, because it's so close, and we can get so much light," she said. "And then, my favorite system ever is TRAPPIST-1, because, seven planets, Earth-size, transiting—it's just the perfect playground."

But the most abundant life-supporting worlds in the galaxy may be duds according to the ease-of-study criterion, lurking in the dark alleys far from the well-lit playground. Over the past few years, astronomers have spotted a fair number of "rogue planets"—worlds zooming alone through the depths of space, unbound to any sun. Such planets are hard to spot, as you might imagine, so discovering even a handful of them suggests that they're incredibly common throughout the Milky

Way. Indeed, many scientists think the rogues far out-number "normal" worlds with a sun. Their profusion isn't all that surprising when you think about it. Gas giants commonly migrate within their natal solar systems, as the abundance of "hot Jupiter" exoplanets shows. These alien gas giants orbit incredibly close to their host stars, with some completing one lap in less than one Earth day. Such worlds couldn't possibly have formed where they now lie, so they must have moved inward—and when they did, they may well have booted their rocky, formerly close-orbiting siblings out into the void, consigning them to an eternity of lonely, benighted wandering.

If these ejected worlds are big enough, they can stay warm and habitable—for microbes, anyway, if not complex and "intelligent" life—for billions of years. Earth's insides, for example, remain molten to this day, thanks to the heat left over from our planet's formation and that given off by the decay of radioactive elements in and around its core.

Indeed, it's possible, Benner said, that "most of the life in this galaxy is on rogue planets that have been ejected from their star."

But finding such life would likely be quite challenging. It's unclear how you'd search a rogue planet's atmosphere for biosignatures without the help of any starlight passing through. We might have to launch tiny probes to explore such worlds up close to get a good idea about their habitability.

Intelligent life may be unmoored not just from stars, but from planets as well. If advanced aliens do commonly transcend biology and go digital, all our notions of "habitable" go out the window. Machine-based ET could be anywhere at all in the depths of space. Indeed, there are reasons to think they'd actively shun planets, whose gravity has a habit of sucking in troublesome comets and asteroids. So, as astronomer Seth Shostak and others have suggested, perhaps SETI searches should target not potentially habitable worlds but rather regions with lots of available energy and raw materials—the stuff machine-ET would need to maintain, repair, and replicate itself—the Milky Way's center, for example, or the coveted real estate on the outskirts of black holes.

Chapter 7

How Would the World Be Told?

Any SETI scientist who gets a promising ping from the great beyond knows what to do next: paint the face of a smiling Grey on 10,000 Ping-Pong balls, lug the lot to the top of the Empire State Building, and drop them all onto the garbage-strewn streets below.

Actually, that's a half-truth. There *is* an established protocol for how to follow up on a detection and communicate a confirmed discovery to the masses, but it differs in several key respects from the one outlined above.

The real instructions, which were adopted by the International Academy of Astronautics (IAA) in 1989, consist of nine rather wordy "principles for disseminating information about the detection of extraterrestrial intelligence."[1] These can be distilled into three basic points.

First, check the signal out to make sure it's the real deal. This is a commonsense measure that probably doesn't need to be in there at all; it's hard to imagine a researcher rushing to announce a discovery before ruling out false positives from orbiting satellites and other sources, which happen all the time in the SETI world. Such confirmation would likely come from a different radio telescope than the one that made the initial detection. If both instruments pick up the same narrow-band signal coming from the same deep-space source, then you're probably clear to proceed to the next step.

Second, don't keep it a secret. The discovery team is supposed to tell the broader SETI community so others can vet it as well. The president or prime minister or junta chief of the nation where the team leader is based should also be looped in—but these bigwigs don't get the honor of announcing the find to the world. That falls to the discoverers themselves, presumably at a press conference that will be fairly well attended.

That last bit may surprise you. Maybe you thought the secretary-general of the United Nations would de-

---

1   You can read the full protocol here: https://iaaseti.org/en/declaration-principles-concerning-activities-following-detection.

liver the news at a special meeting in New York or from the subterranean diamond lair where the global elites meet to discuss alien visitation, the nearly complete plot to create a one-world government controlled by the Freemasons, and other important secret business. But, while the UN would be informed about a detection, the organization has no formally outlined role in spreading the big news.

"We've approached the UN with these protocols," said Seth Shostak, who chaired the IAA's SETI Permanent Committee for 10 years. "They don't have any interest, actually. I think that, if you want to get something really incorporated into the UN's business, you have to have a champion. You have to have some country—you know, Paraguay or somebody has to think, 'Man, this is really important.' Otherwise, they've got other fish to fry."

Indeed, there has been no big push from Paraguay or anybody else on this issue, said Simonetta Di Pippo, director of the UN's Office for Outer Space Affairs.

"The Office for Outer Space Affairs serves United Nations member states, and the office has been given no mandate by member states to develop a protocol or coordination process for discovery of, or contact with, advanced or intelligent extraterrestrial life," she said.

In 2000, astronomers Iván Almár and Jill Tarter proposed a new system for ranking the importance of claimed intelligent alien detections. The metric, called the Rio scale (because the duo first presented it at a conference in Rio de Janeiro), was inspired by the Torino scale, which gauges the risk to Earth posed by incoming comets and asteroids. In case you haven't heard of the Torino scale, let's just say the Rio scale is like the Richter scale, because it kind of is; both of them begin with *Ri*, after all. On the Rio scale, a 0 detection can be dismissed completely, and a 10 signifies "extraordinary importance."

THE RIO SCALE

A few examples would be nice here, and Shostak and Almár helpfully provided some in a 2002 paper. The alien invasion featured in the 1996 film *Independence Day* is a pretty obvious 10, the duo wrote, whereas the black monolith dug up on the moon in *2001: A Space Odyssey* rates a 6 ("noteworthy"). If you'd like something from real life, the so-called face on Mars merited a 2 ("low importance") when first spotted by NASA's Viking 1 orbiter in 1976, but it dropped to a 0 in 2001, after sharper photos by the space agency's Mars Global Surveyor revealed the landform to be just a run-of-the-mill mesa.

The third principle is to not broadcast anything back to our newfound galactic neighbors without "international consultation" (the specifics of which, however, are not spelled out). This is a nod to the debate over METI, or "active SETI," which continues to swirl today. The gist of the controversy, which we'll explore in detail later, is this: some astronomers worry that divulging our existence, location, and PINs to ET could lead to an excruciatingly painful and preventable death by vaporization/nanobot/tentacle slap. Other folks shrug off the risk, reasoning that any aliens capable of doing us harm almost certainly already know that we're here and that we're soft and squishy and vulnerable.

That's the protocol, more or less. But don't get too attached; it's just a set of nonbinding guidelines, toothless as an anteater. There's no international SETI criminal court to try scofflaws for insufficiently rigorous vetting or keeping a junta chief in the dark.

And it's likely that things wouldn't go according to this prescribed plan—specifically, that word would leak out before a promising signal graduates to the level of confirmed discovery. This isn't just a guess; history suggests as much. For example, one evening in June 1997, Tarter, Shostak, and their SETI Institute colleagues picked up something very promising—a powerful narrow-band signal that seemed to be coming from the vicinity of a star about 12 light-years away. For 16 hours or so, the astronomers were cautiously optimistic that this might actually be "the one."

Those hopes faded the next morning, when further analysis traced the signal to the Solar and Heliospheric Observatory, a sun-studying satellite jointly run by the European Space Agency and NASA. But before the team had made that determination, Shostak got a call from William Broad, a reporter at the *New York Times*, asking what was up with the signal. Shostak later learned that Broad had been tipped off via a labyrinthine set of communications involving Tarter; Carl Sagan's widow, Ann Druyan; and the personal assistants for each woman. (You can read all about the 1997 signal in Shostak's 2009 book, *Confessions of an Alien Hunter*.)

This episode and others like it—such as that possible signal flagged in May 2015 by Russian astronomers—are instructive in a few different ways. For example, they suggest that a true SETI discovery would be a halting, stepwise process, not a single glorious "Eureka!" moment.

And they also put the lie to conspiracist notions that a *Men in Black*–style special-ops team will swoop in once a find is made, zap everybody with a neuralyzer, and haul all the relevant computers, documents, and corpses off to a meat locker in Area 51.

"The government didn't have any interest, of course," Shostak said of the 1997 signal. "It wasn't the government that was calling; it was the *New York Times*."

Of course, if the detection came just ahead of an invasion, the government would be very interested. And you'd probably find out pretty quickly, via an air-raid siren and emergency Grey Alert on your phone.

This protocol we've been talking about applies only to SETI—the hunt for advanced or intelligent extraterrestrials. There are no similar instructions regarding the discovery of "simple" life, the kind we might find in a Martian meteorite or via a long-distance look at TRAPPIST-1f's atmosphere. These latter finds would probably just go through the normal scientific process of peer review, publication in a major journal, and announcement via news conferences and press releases, though there would of course be quite a bit more fanfare for a purported ET discovery than for a new estimate of the mass of the monster black hole lurking at the Milky Way's heart. (Not that this wouldn't be interesting. Astronomers think this black hole, known as Sagittarius A*, contains as much mass as 4 million suns. But that's relatively small potatoes for a supermassive black hole: some weigh in at 10 billion solar masses or more.)

"In a way, it follows this path that we've been treading," Lisa Kaltenegger, director of Cornell's Carl Sagan Institute, said. "They found the first [alien] planets—that was a normal scientific news conference. They found the first planet in the habitable zone—that was a normal scientific discovery. Then they found the first potentially small one, the first potentially habitable planet in the habitable zone—that was a big news conference."

Again, we can look to history as a guide—the 1996 Allan Hills meteorite alien-life claim. The researchers behind that study published their stunning results in the prestigious journal *Science* and announced it to the broader world via a press conference that involved then president Bill Clinton. Fanfare indeed. But, more than two decades later, astrobiologists are still arguing about that Red Planet rock, as we shall see.

Chapter 8

Could We Talk to ET?

If we stumble upon a random ping from ET, one that wasn't meant for us, we'll have an awfully hard time dredging any meaning out of it.

Remember: any aliens capable of transmitting signals into the great beyond are almost certainly far more advanced than we are and different in fundamental but unknowable ways. For example, what if the creatures responsible for our make-believe missive perceive their environment primarily via touch or smell rather than sight? A wave of Smell-O-Vision signals from Planet Grey could wash over us a thousand times and would probably still go right over our heads.

If you're skeptical of this claim, consider that many ancient writing systems here on Earth remain undeciphered today, despite decades of effort by anthropologists and linguists. And these were all crafted by our fellow humans.

"I think if it's just kind of an artifact—if it's the equivalent of our television—we will know they're out there, but we'll probably never be able to figure out the content of that kind of signal," SETI scientist Dan Werthimer said. "So, we'll know we're not alone, but we won't really know that much about their civilization."

This doesn't mean we shouldn't try. If we did get such a hit, SETI scientists all over the world would study the hell out of it with the scopes they have—assuming the signal was constant, and not some sort of one-off event like the Cubs winning the World Series—and they'd probably clamor for a gigantic new instrument to look at the signal even more closely. (In *Confessions of an Alien Hunter*, Seth Shostak notes that decoding a TV broadcast requires about 10,000 times more signal sensitivity than that needed just to pick up the ping. A dish designed to sift an ET signal for meaning would therefore have to be many miles across to get the job done.) I bet they'd get the money to build it, too. Most national governments would be quite keen to get dibs on any info that astronomers manage to ferret out.

Hmmm. Are you now getting visions of North Korea building a miniature Death Star inside a hollowed-out mountain near the Chinese border? If so, don't worry—

you're not (necessarily) crazy. Other folks worry about this stuff, too.

"If we could magically tap into an understanding of science and a consequent development of technology that is thousands or millions of years ahead of where we actually are—given the bad outcomes that might result from just one person abusing that, maybe it's a good thing that we wouldn't be able to decipher it," said Nick Pope, who worked for the British Ministry of Defence for 21 years and ran its UFO investigation project from 1991 to 1994.

REACHING OUT TO US

It's a different story altogether if the ping we get is a deliberate attempt to communicate. Then, we're in with a chance—a very good chance, in Werthimer's eyes.

"Maybe they are interested in young, emerging civilizations like us, and maybe they want to help us get on the galactic internet or something, and they would transmit a signal our way," he said.

Indeed, it wouldn't be too tough for advanced aliens to determine that Earth hosts life (or, at least, that it's extremely likely to do so). Even if they're too far away to have experienced our TV and radio leakage yet, ET scientists could theoretically still detect oxygen, methane, and other biosignature gases in our atmosphere.

"If it's something like that, I think it'll probably be very simple to decipher," Werthimer added. "It'll have language lessons, be anti-cryptographic—you know, lots of pictures."

This is how we would do it, anyway—and how we actually have done it, on a few occasions. In 1974, for example, a team that included Frank Drake and Carl Sagan used the huge Arecibo Observatory in Puerto Rico to beam a three-minute radio missive to M13, a globular cluster 25,000 light-years away. The famous Arecibo Message contained pictures—graphical representations of a person, our solar system, and the observatory itself, all of them super-pixelated, 1970s-style. Google the image: the whole decoded thing looks like a hidden bonus level of Space Invaders.

But that's not all. The Arecibo Message also featured the molecular formulas of some key constituents of DNA, as well as the atomic numbers of the elements most essential to life on Earth—carbon, hydrogen, nitrogen, oxygen, and phosphorus. And it encoded the numbers one through ten. All of this information was packed into 1,679 bits, which was no accident: 1,679 is the product of two prime numbers, 23 and 73. In fact, those are the only two whole numbers you can multiply to get 1,679.

This little Easter egg is a nod to a notion shared by many astronomers—that math is a more or less universal language throughout the cosmos, and that any aliens capable of transmitting and receiving messages will therefore speak it in roughly the same way we do. This idea has been around for a long time. Back in the early 1800s, for example, our friend Carl Gauss purportedly proposed carving a gigantic right triangle into the Siberian forest, thereby showing any Martians or lunarians who might be watching that we are here, and we know all about the Pythagorean theorem. (This famous right-triangle formula—$a^2 + b^2 = c^2$—would hopefully impress ET more than it impresses potential prom dates.) The huge landscape architecture project never got off the drawing board, however.

The Pythagorean theorem was also part of the message that a team led by Douglas Vakoch of METI International beamed in October 2017 to the potentially habitable exoplanet GJ 273b, which lies about 12 light-years from Earth. The astronomers built up to this little trigonometry lesson, starting the signal out with simple tutorials on counting and arithmetic. They also sent the message three separate times, so ET won't be stymied by any glitches that may sneak in during the long journey through space: If there's conflicting information, Mr. Grey, just go with the version that occurs two out of three times. (As these examples show, there is no purple-robed authority figure who speaks for Earth in our dealings with ET. If you have access to a radio telescope, you can assume this responsibility for yourself.)

Maybe ET will use math to get our attention—a string of prime numbers, perhaps, or the apparently omnipresent, omni-important Pythagorean theorem. High school trig teachers should rebrand it the alien communication formula. That would've kept you from daydreaming in class, right?

THE ARCHAEOLOGY OF THE FUTURE

We'd have to rely on our own smarts to decode any ping we get from ET; the vast scale of the cosmos makes asking the senders for help a no-go. If we were puzzled by a signal beamed out by the lords and masters of GJ 273b, for example, it'd take 12 years for them to receive our head-scratching and shruggy-guy emoticons, and another 12 years for us to get their (perhaps unintelligible) reply.

And in many cases, the signal we get may have come from a civilization that died out long ago. Think about the Arecibo Message: what are the odds that humanity will still be around in the year 26,974, when that Atari-esque stream streaks through the M13 system? How about in 51,974, when we could expect a reply? (Actually, the chances of a response from M13-ians are even lower than you might imagine: the senders beamed the message to where the cluster is now, not where it will be in 25,000 years.)

This is why SETI pioneer Philip Morrison—who, along with Giuseppe Cocconi, wrote the field's founding document, a 1959 paper laying out why scientists should search for radio signals from alien civilizations—described SETI as "the archaeology of the future." Indeed, our interactions with advanced aliens may end up being similar to our dealings with the ancient Greeks and Romans, Jill Tarter said.

"They have transferred information forward in time that we can read today, and understand, and learn, even though we never got to ask them questions," she said. "So, that might be the appropriate model for interstellar communication, if there is in fact information embedded in the signal and we can figure out how to decode it. We could still learn a lot."

But don't get the wrong idea: not everyone shares Werthimer's optimism that we'd be up to the decoding challenge. "I think our notion that we could interpret a SETI signal if one were to be detected is actually a very funny one," Lisa Kaltenegger said. "I can't communicate with a jellyfish, let alone if I don't see it. And that evolved on our own planet."

FACE-TO-FACE

You may be thinking, "Hey! You haven't talked about an alien arrival, like in that alien movie *Arrival*." You're right. So let's do that, briefly.

Why only briefly? Because this is a much less likely scenario than picking up a ping with a radio telescope. It's far easier, cheaper, and faster to send electromagnetic signals across the galaxy than a starship packed with people-probing tools, after all.

It stands to reason that we'd have a much better shot at understanding ET if he/she were standing right in front of us. As long as the aliens perceived reality in the same way that we do—saw light in the same wavelengths, for example—we could theoretically do enlightening little soft-shoe routines for each other, as depicted in *Arrival*.

But still, there's no guarantee that we could communicate in any deep and meaningful way across the vast gulf carved out by our divergent origins, evolution, and technological capability. Say, for example, that the heptapods in *Arrival* really do show up here tomorrow, and they start grinding us into pink goo to feed their nonapod pets. Could we negotiate our way to salvation?

Shostak likened our chances in such a scenario to those of dinosaurs facing down time-traveling human hunters determined to put their scaly heads on a wall.

"I think you run," he said. "Just head for the hills—that's probably the best strategy."

Chapter 9

How Would the World Respond?

Chris McKay got into the astrobiology game to help answer the pesky "Are we alone?" question. A bona fide ET detection—by him or by anybody else—would satisfy his curiosity and therefore spur a career change: "I'd go back to repairing motorcycles for a living," he said with a laugh.

But how would the rest of us deal with the news? What would society's reaction be? Let's take a few stabs.

MICROBES VERSUS GREYS

There are a lot of variables to consider, the first of which is what kind of alien we manage to dig up. We're going to start with microbes—or, rather, "simple" life-forms in general—because they're the most likely catch. After all, these little guys may well be in our own backyard, swimming in the seas of Enceladus and Europa or scratching out a meager living deep beneath Mars's shifting red sands.

We already know broadly how this would go, thanks to a famous test case—the Allan Hills 84001 saga. The detection of purported signs of life in this Mars meteorite was a big deal back in 1996; it made front pages of newspapers all over the world, and US president Bill Clinton even gave a speech to mark the occasion.

"You didn't see massive social upheaval as a result,"

said Michael Varnum, a psychologist at Arizona State University. "A lot of people thought it was neat."

And that reaction would probably hold today, as some recent research by Varnum and his colleagues suggests. In 2017, the scientists presented a (slightly modified) 1996 *New York Times* article about ALH 84001 to survey participants as a new story. Folks responded much more positively than negatively to the "news"—which makes a lot of sense, considering how we tick.

"We know that part of human psychology is a desire for, and an appreciation of, novelty. It's why you go over the next hill, or you hop in a raft and cross an ocean," Varnum said. "And this is about as novel as something could be: lifeforms from outside of the planet."

He did add an important caveat, however: "As long as that novelty is nonthreatening—you know, it's not pointing a giant space laser at you."

Indeed. It's safe to assume that the discovery of intelligent aliens would freak us out, if the creatures in question were piloting an armada laying waste to Earth and beaming our fleeing forms into the subterranean salt mines of GJ 273b.

Even a mere SETI signal could be disruptive, if we manage to decode it. Say, for example, ET's message contains a blueprint for that mini Death Star we talked about in chapter 8, or an antiaging recipe that, unlike all those $200 skin creams, actually works. Different groups here on Earth might fight over access to such advanced technology or use it in ways that most of us wouldn't appreciate.

"You could imagine, if the militaries of the world got ahold of some advanced technologies, that could be really bad," Werthimer said.

But a mysterious, abstract ping would be a much different story. The signal by itself, with no follow-up—and it's entirely possible that there would be no meaningful follow-up on a SETI detection, ever—would likely elicit a reaction similar to that greeting Mars microbes or minuscule Enceladusians. The shock quotient in such a case would be low, because many of us have already concluded we're not alone. Polling repeatedly shows that about half of Americans believe intelligent aliens exist— and 30 to 40 percent of us even think they've already visited Earth.

"It'll be front-page news for a little while, but then the Kardashians will do something, and the media will go chase that," Tarter said.

This prediction also accords well with our knowledge of human nature. We tend not to get too worked up about things that don't directly affect us or our loved ones now or threaten to do so in the immediate future.

THE LONG HAUL

So far we've considered just the short-term impacts of a detection, which would probably be limited to a temporary and mild endorphin rush and perhaps the purchase of a telescope that gets used twice and then stowed, forgotten, in a closet. (There are exceptions, of course. NASA's budget would almost certainly get a boost, and astrobiologists would have lots of new research avenues to explore.)

But there would be long-term effects as well, as humanity slowly comes to grips with the reality that the cosmos is brimming with life.

"It basically makes us question everything we know about ourselves and our place in the universe," Zurbuchen said. "It's inescapable that that question needs to be asked. And these things are often not scientific discussions; they're asked in many different realms. Philosophers talk about it; religious leaders talk about it. And I think all these discussions will be relevant."

Indeed, the discovery of alien life may eventually spur a reassessment of ourselves as profound as those forced by the Copernican and Darwinian revolutions, which showed that (1) we're not the center of the solar system, let alone the universe, and (2) our ancestry, like that of all creatures on the planet, goes all the way back to the primeval muck.

The impact will be bigger, of course, if the epochal detection is a SETI signal rather than a fossilized Martian microbe or fish-like fingerling on Europa. Super-smart aliens would hit hard at our sense of specialness and superiority, just as Copernicus and Darwin did.

That's not necessarily a bad thing.

"There's a humbling that could happen," said Jonathon Keats, a conceptual artist and experimental philosopher who thinks a lot about humanity's place in the universe. "That's the best possible outcome."

Are you wondering about religion? Well, it survived Copernicus and Darwin, and the potential Alien Revolution[1]—even the SETI variant—isn't likely to reduce churches, mosques, and synagogues to rubble, either.

This isn't just arm-waving; we have some data.

In 2008, theologian Ted Peters surveyed more than 1,300 people representing a variety of faith systems, from Catholics and Protestants to Jews, Mormons, Buddhists, and the nonreligious. He asked these folks, among other things, whether they agreed or disagreed with the follow-

---

1   *Alien: Revolution* will be part of the famous movie franchise someday, probably a year or so after *Alien: Taxation without Representation* hits theaters.

ing statement: "Official confirmation of the discovery of a civilization of intelligent beings living on another planet would so undercut my beliefs that my beliefs would face a crisis."

The overwhelming majority of people disagreed, and this sentiment was unanimous across faith groups. In fact, at least 83 percent of respondents in every category either disagreed or strongly disagreed with the statement.

"It's my own belief, both theologically and according to this survey, that we're not likely to see a crisis in religious belief in almost any religion," said Peters, who's based at the Pacific Lutheran Theological Seminary in Berkeley, California.

That being said, some religions do seem more vulnerable to a shakeup than others. For example, Buddhists and Mormons would almost certainly take the news in stride, while Muslims and non-Mormon Christians would probably have a harder time, Peters said.

"The Mormons already have a doctrine of extraterrestrial life, and the Buddhists don't seem to be bothered by anything in the physical world," he said. "The Christians and the Muslims really want to know about what it all means, and integrating an entire off-Earth history into the story of salvation will be a lot harder for them."

Fundamentalist Christians, who view the Bible as the revealed and complete word of God, would likely face the most serious internal struggles, said astronomer David Weintraub of Vanderbilt University, author of the 2014 book *Religions and Extraterrestrial Life: How Will We Deal with It?*

"The very fundamentalist Christian writers say this on their websites and in their writings—there is no life out there, because if there were, God would have written about it and told us about that in the Bible," Weintraub said. "So those folks would be surprised and bothered, but I don't think anybody else would."

Some Christians may also be bothered by another issue: does God save the souls of intelligent aliens? If so, does He/She manifest in billions of different alien forms on billions of planets throughout the cosmos? Or was there just the one incarnation (Jesus), especially for us?

The appraisal of the ALH 84001 evidence played out over years—it's still going on, as a matter of fact—and a similarly lengthy vetting process would likely accompany any putative alien-life find before the scientific community fully accepted it. People around the world, religious or not, would probably have plenty of time to get used to the idea of a discovery before it's official, further reducing the chances of upheaval, Weintraub added.

Coming to grips with a new view of the cosmos and our place in it may end up being a natural part of mainstream

Christianity's maturation process. Learning that their father has a life outside the house shouldn't make Christians feel any less special, said Brother Guy Consolmagno, director of the Vatican Observatory.

"If anything, it reflects better on us that our daddy is the daddy who did all of this," Consolmagno said. "Finding that there are more civilizations out there— that there's even more amazing things that our daddy did—shouldn't make us feel smaller. It should make us feel all the more amazed that the creator of these multiverses, or whatever they are, also has time to pay attention to us."

One more quick note before we move on: I think it's a safe bet that the discovery of alien life, especially intelligent alien life, would spur a rash of religious weirdness that makes the 2012 Mayan apocalypse craze look as measured as Murrow. Alien-worshipping cults will pop up around the world, predictions about the Second Coming will ramp up in volume and stridency, and folks will start seeing the Virgin Mary in Martian rock formations photographed by the Curiosity rover. (More than they do already, that is. Some people spend hours and hours poring over Curiosity's archived photos online, searching for signs of alien life that NASA is hiding from us. Some purported finds include a Mars rat, lizard, and crab.) That sort of thing. It'll be like the Second

Great Awakening, but with aliens and fiery services led by weirdos on YouTube instead of in clearings in the woods.

Think about Heaven's Gate: 39 members of this cult killed themselves in 1997 in an attempt to board an alien spacecraft flying behind Comet Hale-Bopp. And that starship was entirely make-believe.

FAKE NEWS?

Throughout this entire discussion, we've been blithely assuming that everyone will accept the discovery as real. This is obviously not the case. After all, we live in a country where millions of people believe NASA faked the Apollo moon landings, and an unknown but disturbingly high number think the space agency, in league with an international cabal of smug elites, has fooled most of us sheeple into believing the Earth is round. (One is a disturbingly high number when it comes to this flat Earth nonsense, which is the stupidest thing in the whole wide world, and the flat Earthers are numerous enough to hold conferences; their first international meeting took place in November 2017, in North Carolina.)

The same folks who think NASA fabricates every

Earth-from-space-photo will cry "Hoax!" about a narrow-band SETI signal from M13 and airily dismiss a spectrum showing oxygen and methane in an exoplanet's atmosphere. A photo of a Mars microbe could be doctored—maybe it's just a regular old Earth bacterium in a red wig.

Some of the doubt will be faith-based, coming from the same wounded place that denies Darwin because "I'm not descended from a damn monkey." And some of it will stem from a distrust of authority, the pleasure that comes from feeling smarter and more enlightened than the masses, the prioritization of personal experience over fancy book learning, and all the other reasons people believe in conspiracy theories.

How many voices will raise that "Bullshit!" cry? Will they drown out the approving little grunts of more-than-mild interest that the rest of us are making at the same time? Your guess is as good as mine.

A SETI detection "has great potential to help us overcome a lot of our differences," Keats said. "But doubt and skepticism could also result in a bad divide, a schism between believers and nonbelievers."

Chapter 10

Have We Already Found ET?

Gil Levin thinks so.

Levin was the principal investigator for the Labeled Release experiment, part of the life-hunting science package that NASA's Viking 1 and Viking 2 landers toted to Mars in the mid-1970s.

Labeled Release fed drops of water onto little scoops of Martian soil. These drops were basically spoonfuls of microbe stew; they were chock-full of amino acids and other organic molecules that any tiny Martians lurking in the red dirt could gobble up. If this free lunch were accepted, Labeled Release would know; those nutrients were specially branded with radioactive carbon-14, which an onboard Geiger counter would detect in the microbes' gaseous metabolic products.

That did indeed happen—or so it seemed at first. The Vikings' Geiger counters picked up a healthy stream of radioactive carbon dioxide bubbling out of the dirt. (If any stream of radioactive CO_2 can be considered healthy. You wouldn't want to swim in one, or eat a fish pulled from its froth.) Heat-sterilized control samples showed far less activity, which was also great news for the we-are-not-alone crowd: dead microbes tell no tales and sup no stew. Slam dunk, right?

Well, not exactly. The two other life-detection experiments aboard the Vikings returned negative or ambiguous results. Most damningly, the landers apparently found no native organic molecules in the red dirt—just a couple of weird chlorine-containing compounds that mission scientists figured were contaminants brought from Earth. (Specifically, residues left by the fluids used to clean the Vikings' instruments before launch.) Martian soil appeared to be even more carbon-free than the dead gray dust on the surface of the moon. You can't have life as we know it without organics, so a consensus quickly emerged that what the Vikings had found was probably evidence of exotic Martian chemistry rather than alien biology.

But that consensus has crumbled a bit over the past decade. The key moment came in 2008, when NASA's Phoenix lander detected chlorine molecules called perchlorates in the soil near the Martian north pole. Lab experiments here on Earth soon showed that, if you heat up perchlorate-rich soil—as the Vikings' organics-hunting instrument did, to drive off the molecules hidden within—carbon compounds get "burned" into CO_2, chloromethane, and dichloromethane. These latter two are the very same chlorine molecules the Vikings saw on Mars.

So, do the building blocks of life knock about in the Martian dirt? Did the Vikings actually spot signs of alien metabolism? Nobody knows for sure, and scientists are debating the various interpretations with renewed vigor. But through all the twists and turns, Levin has been un-wavering. He has said all along that Labeled Release found evidence of Mars life, and he still feels that way.

"More strongly than ever," Levin said. "Ever since I first uttered that conclusion, all new data have been consistent with it or supported it. There have been no in-imical findings at all."

IT CAME FROM MARS

Twenty years after Viking, the world got zapped by an-other rust-colored lightning bolt.

In August 1996, a team of scientists led by David McKay of NASA's Johnson Space Center announced some big news about Allan Hills 84001, a 4-pound meteorite that had been scooped up in the Allan Hills region of Antarctica in 1984. The southernmost continent is a great hunting ground for space rocks, by the way, because it's so barren; meteorites stand out on its frosty white surface like nuggets of gray-black gold.

ALH 84001 formed in the very distant past, not too long after Mars itself took shape, and was blasted into space by a powerful impact 17 million years ago. The rock zoomed around the sun a few million times before getting sucked in by Earth's gravity 13,000 years ago, about the same time the first pioneering humans reached North America.

Entombed within this interplanetary time capsule, McKay and his colleagues reported, were carbonate minerals, which speak of the long-ago presence of liquid water; organic molecules called polycyclic aromatic hydrocarbons; tiny spherical and tube-like structures that look a lot like fossilized microbes (most famously, the "worm" that pops up first in every ALH 84001 Google Image search); and minuscule crystals of pure magnetite resembling those produced by bacteria here on Earth.

"Although there are alternative explanations for each of these phenomena taken individually, when considered collectively, particularly in view of their spatial association, we conclude that they are evidence for primitive life on early Mars," the researchers wrote in their discovery paper, which was published in the journal *Science*.

Wow. Yes. This was a very big deal. President Bill Clinton gave a brief speech about ALH 84001 on the White House lawn on August 7, 1996, a few minutes before McKay and his team officially announced the news to the masses during a press conference at NASA headquarters. In the days leading up to the big reveal, Clinton had discussed the meteorite and its potential importance quite a bit with his aides, including Dick Morris, who ended up spilling the beans to a Virginia call girl named Sherry Rowlands. We know this delightful little nugget because parts of Rowlands's diary made it out into the world, and the entry for August 2, 1996, includes the following: "He said they found proof of life on Pluto!" (You can learn more about this and all the other drama surrounding the ALH 84001 announcement in Joel Achenbach's 1999 book *Captured by Aliens* and Kathy Sawyer's 2006 book *The Rock from Mars*.)

But that was far from the end of the story. As had happened two decades earlier, other researchers soon started trying to poke holes in the Mars life interpretation. And some of them did so with an aggressiveness and lack of civility that caught McKay and his colleagues off-guard.

"I wish someone had told us what was coming down the

pike," said team member Kathie Thomas-Keprta, who's also based at JSC. "It was over-the-top contentious."

The organic molecules and the magnetite crystals could have been produced nonbiologically, some doubters said. Others suggested that the possible "microfossils" may just be residues of the coating used to prepare ALH 84001 samples for their electron-microscope close-ups. Some of those structures are too small to be life as we know it anyway, yet others said. After all, cells have to be big enough to fit all the molecular machinery that keeps us alive, such as ribosomes, which synthesize proteins by linking amino acids together. Some of the ALH 84001 structures were about 20 nanometers wide—about the same size as a single bacterial ribosome. (Others were much bigger, however, and fell comfortably within the size range of Earth microbes.)

The debate over the Mars rock rages on today. Some researchers—including Thomas-Keprta and fellow discovery team member Everett Gibson, also of JSC, as well as astrobiologist Dirk Schulze-Makuch—continue to argue that ALH 84001 harbors solid evidence of alien life.

"If you put it all together, it makes sense," Schulze-Makuch said. He thinks much of the opposition to the 1996 paper stems from a public relations issue: NASA, he said, focused too much on the "microfossils," presumably because they were so photogenic, even though the magnetite crystals were, and remain, the strongest line of evidence.

A larger proportion, however, remains unconvinced.

"Everybody wanted it to be life. Everybody wanted evidence of life, so nobody paid any attention to the caveats," said Chris McKay (no relation to David McKay, who died in 2013 at the age of 76). "And it's not just the scientists, but the whole world, in a sense. Everybody wants it, and I do, too—I want there to be life on Enceladus and Mars. I wouldn't be in this field otherwise. So, it's easy to get taken up by that and set aside your critical training."

Is there a takeaway from these twin tales? Yes, a few. One of them Chris McKay summed up pretty nicely right there. And another we discussed in chapter 5: the bar is very high for a widely accepted detection of alien life. Carl Sagan's famous phrase gets trotted out a lot, but it really does apply here: "Extraordinary claims require extraordinary evidence." We're talking about the most dramatic and important discovery in the history of humanity, apart from the revelation that Jon Snow is the son of Lyanna Stark and Rhaegar Targaryen, so the evidence will have to be pretty much bulletproof. (And even then, many people doubtless will choose not to accept it, as we discussed in chapter 9.)

One more thing before we move on: however you feel about the evidence David McKay and his team marshaled, the ALH 84001 study had a huge impact on planetary science, and Mars exploration in particular. "I am determined that the American space program will put its full intellectual power and technological prowess behind the search for further evidence of life on Mars," President Clinton said during that speech on the White House lawn. Now, that statement didn't turn out to be 100 percent true. After Viking, NASA's Mars ambitions shifted from life detection to an investigation of Mars's *potential* to host life—basically, an admission that we don't know enough about how the Red Planet works to mount an effective search for microbes there. That habitability focus remains today (though the Mars 2020 rover mission will mark a return to hunting for life, albeit the long-dead kind.)

But NASA did ramp up its robotic Mars program in the Allan Hills aftermath, and you can still see the effects today. As of this writing, the agency has five spacecraft actively studying the Red Planet, both on the surface and from orbit—the Opportunity and Curiosity rovers, the Mars Odyssey orbiter, the Mars Reconnaissance Orbiter, and the Mars Atmosphere and Volatile Evolution probe (which is also an orbiter). And a lander called InSight launched toward the Red

Planet in May 2018; it's scheduled to touch down in late November.

"It jump-started, you might say, the whole program of planetary exploration to Mars, and the field of astrobiology," Gibson said. "The administrator of NASA, Dan Goldin, told us, 'You know, you guys realize, you put $6 billion into the planetary exploration program, saved JPL [the Jet Propulsion Laboratory in Pasadena, California, the agency's lead center for robotic exploration missions to other worlds] 2,000 jobs, and now what you have created and we've created with the astrobiology institutes and the interdisciplinary projects that are going on—that's a contribution that you all should be proud of.'"

UFOS: INTELLIGENT ALIENS?

You've probably seen the video: a glowing blob skims above the clouds, framed by the crosshairs of a navy jet's infrared camera. The blob rolls and rotates, eliciting excited and incredulous chatter from the airmen tracking it. "Look at that thing, dude!" one of them exclaims over his radio.

This UFO encounter, which took place in 2004 about 80 miles off the San Diego coast, was one of many investigated by a Pentagon project called the Advanced

Aviation Threat Identification Program. AATIP operated from 2007 through at least 2012, but we learned about it only in December 2017, via near-simultaneous bombshell reports in the *New York Times* and Politico.

And they were bombshells. The US military had officially been out of the UFO investigation business since the shuttering of the Air Force's famous Project Blue Book, which checked out more than 12,000 reported sightings between 1952 and 1969. Why wade back in with AATIP? Why not tell us about it? And just what did the program find?

As of the time of this writing, we still don't know the answers to these questions, at least not fully. What we do know, according to the *Times* and Politico stories: AATIP was championed by former Senate majority leader Harry Reid, a Nevada Democrat; it apparently cost taxpayers a total of $22 million over its run; much of this money went to Las Vegas company Bigelow Aerospace, which is headed by entrepreneur Robert Bigelow, a UFO enthusiast and Reid's longtime friend; and, last but certainly not least, Bigelow Aerospace has apparently modified some of its buildings to store "metal alloys" that AATIP honchos say were recovered from UFOs. (Another semantics note: believers often use the term *unidentified aerial phenomena* as a way to avoid the tinfoil-hat stigma that has become associated with *UFO* over the years.

And when I say "believer," I'm referring to people who think that some observed UFOs are alien spacecraft. The existence of UFOs in the broader sense is irrefutable: people do see things in the sky that they can't identify.)

As a big fan of alloys in general, and alien alloys in particular, I would really like to learn more about this last bit. I'm sure you would, too. But so far, the purported UFO alloys remain locked away in those mysterious modified buildings in the Nevada desert.

The AATIP revelations have doubtless confirmed the suspicions of many UFO believers—namely, that intelligent aliens have journeyed to Earth many times, and the government knows about these visits but is keeping the juicy news from us. There are many such believers out there, as we noted in chapter 9. One of them is Jan Harzan, executive director of the Mutual UFO Network, who offered up two possible reasons for the ongoing cover-up.

First of all, he said, the government is worried that we can't handle the truth—that it would be too disruptive to the world economy and society in general.

"The second is that they don't want our enemies getting ahold of this technology," Harzan said. "If you can manipulate space-time, or you can move things around at light speed—I mean, Saddam Hussein or our little friend Kim over in Korea—you could put a nuclear bomb on the doorstep of the White House instantaneously and blow it up before anyone had any time to react to it."

The skeptics, of course, see things very differently. They tend to note, for example, that it's perfectly natural for the US government—indeed, any government—to be interested in weird objects zooming through its airspace. In fact, you might consider your leaders incompetent if they didn't check this stuff out, given all the high-tech spying that countries have gotten up to over the years.

Also, the Pentagon stopped funding Project Blue Book and AATIP, with the stated justification that these programs didn't turn up any proof of alien spacecraft—or anything else terribly interesting. This basic finding applies to the entire catalog of UFO sightings,[1] said retired air force pilot James McGaha, who's now an astronomer and director of the Grasslands Observatory in Arizona.

"There's absolutely no evidence," said McGaha, who

---

1   This catalog is pretty big: there have been more than 100,000 reported sightings since 1905. A hugely disproportionate share come from the US, either because ET is most interested in the world's foremost superpower or because we're the most steeped in UFO mythology, depending on your worldview. See https://vizthis.wordpress.com/2017/02/21/i-want-to-believe-ufo-sightings-around-the-world.

has investigated and evaluated many UFO sightings over the years. "None. Never has been."

Take the most famous UFO incident of all time, for example. There's no evidence that the US government found a crashed alien spacecraft near Roswell, New Mexico, in 1947 and dragged its bulbous-headed, long-fingered passengers to Area 51 for dissection. Instead, everything points to the crash of a balloon system that was monitoring the atmosphere for signs of Soviet nuclear tests, as part of the military's then-top-secret Project Mogul, McGaha said. (The Roswell incident faded into obscurity for three decades, he added, only surging back into the public mind in 1978 when some UFO enthusiasts began investigating and promoting it.)

That dramatic 2004 video we discussed above? The glowing blob looks an awful lot like the heat signature of a distant jet, McGaha and other skeptics have said. (Remember, that footage was taken with an infrared sensor.) Indeed, many of the most puzzling and suggestive UFO incidents may be traceable to aircraft, especially stuff we're not supposed to know about, according to Nick Pope, who investigated UFO sightings for the British Ministry of Defence.

"Aviation technology probably runs 10, 15, some say 20 years ahead of what's been publicly declared," he said. "If it's a good sighting with, say, credible witnesses and even something additional, like photographs, video, or radar, I think just statistically it's far more likely that we're dealing with some deep-black project—some secret prototype aircraft or drone."

In many cases, we may never know. For instance, about 700 of the sightings investigated during Project Blue Book remain unresolved. That's frustrating. It sucks. But just because we haven't figured out what a UFO is doesn't mean it's an alien spacecraft. And that's an unwarranted logical leap that too many people are ready to make.

That leap is so tempting. We all want to make it, at least subconsciously. We've been primed to accept the existence of ET, both from science fiction and from all the exciting and mysterious discoveries we've been talking about in this book—possible signs of life on Mars, buried oceans in the outer solar system, potentially Earth-like alien planets, weirdly dimming stars that get even scientists saying the words *alien megastructure*, strange radio pings from space, and so on. A part of us is very ready to believe, proof or no proof. Polls consistently show that about half of Americans think ghosts and angels are real, for example.

McGaha sees that connection, too.

"It's a belief system," he said. "It's a myth; it's a superstition. It makes people feel good."

This same basic argument applies to alien abduction stories as well. Skeptics have noted that many purported abductees recall their experiences only years later, often under hypnosis, and that incidents often occur when a person is in the fuzzy gray zone between sleep and wakefulness, reality and dream. Michael Shermer makes these points in his 1997 book *Why People Believe Weird Things: Pseudoscience, Superstition, and Other Confusions of Our Time.*

As you can tell by now, I put myself squarely in the skeptic camp. But that doesn't mean I regard the UFO believers as stupid or naïve; I just don't think anyone has made a compelling case for visitation yet. Extraordinary claims, and so forth. My gut tells me that ET's out there, somewhere—but it also tells me that the odds of a visit during my brief time on this planet aren't good, and that, if it happened, we'd probably all know about it.

Chapter 11

Will Aliens Kill Us All?

The sun peeks over a hill, casting slanting golden light onto a rock-strewn meadow near the eastern shores of Lake Ontario. Fog hangs low over the grass, lower even than the surrounding trees, but in shreds too ragged and wispy to obscure the riot of humanity roiling there—hundreds of nearly naked, stick-wielding men chasing a tiny ball made of deerhide.

That's what's happening in one little patch of North America on the morning of October 12, 1492: a rousing game of baggataway (a pastime that eventually morphed into lacrosse). Two thousand miles to the southwest, in the Salt River Valley of present-day Arizona, Akimel O'odham farmers tend to their beans, corn, and melons. Near what is now Mexico City, an Aztec priest rips the still-beating heart out of a captured enemy warrior in a temple-top ceremony, then kicks the bloody body down the ziggurat's wide stone steps. (I don't mean to imply that human sacrifice defined the rich and complex Aztec culture. But, hey, if it bleeds, it leads.) And on a small island in the Bahamas, three small ships sailing under the Spanish flag come ashore, ending a nine-week journey across the Atlantic.

That landfall changed everything. Just three generations later, the Spanish had overthrown the most powerful empires the New World had ever known. The Aztecs went down in 1521, while the Incas, who once ruled lands from central Chile all the way up into Colombia, fell in 1572. Millions of native peoples throughout the region died or were thrown into slavery, forced to swing pickaxes in gold and silver mines or scythes on sweltering sugarcane plantations.

Stephen Hawking viewed this sad history as a cautionary tale. If intelligent extraterrestrials exist somewhere out there in the Milky Way, all of humanity could suffer the fate that befell the Aztecs and Incas, the cosmologist repeatedly warned.

"Such advanced aliens would perhaps become nomads, looking to conquer and colonize whatever planets they could reach," Hawking said on a 2010 episode of *Into the Universe with Stephen Hawking,* a TV show that aired on the Discovery Channel. "If so, it makes sense for them to exploit each new planet for material to build more spaceships so they could move on. Who knows what the limits would be?"

Again in 2016, in the documentary *Stephen Hawking's Favorite Places,* he said, "Meeting an advanced civilization could be like Native Americans encountering Columbus. That didn't turn out so well."

To be fair, such interlopers could be driven by need rather than greed. Maybe some Vader-like galactic overlord destroyed their home planet, for example, and they hope to rebuild their civilization on Earth, the most promising oasis for light-years around. There's just the small matter of ridding this new world of its previous rulers, a strange race of mostly hairless bipeds that steal and drink the nourishing liquid that other, hairier species produce for their own babies.

And if you believe humanity would band together and valiantly fight off the tentacle-waggling invaders, you've been watching too many alien-punching sci-fi films—like *Independence Day,* in which Will Smith's fighter pilot character literally punches an alien in the face. Depressingly, we might even fall at the "band together" hurdle, Ronald Reagan's sunny optimism notwithstanding.[1] After all, we've already faced several existential threats already—nuclear holocaust and the ongoing climate change crisis spring to mind—and have so far squabbled and bickered our way to contested, tenuous half solutions. Selfish tribalism seems to be a defining part of human nature, like the capacity for complex abstract thought, a nonsexual love of Tom Hanks, and a crippling fear of clowns.

But say we do buck the odds and all join hands against ET. There's still the fighting—or, rather, the cowering, the running away, the surrendering, the getting melted by ray guns or acid blasters or whatever super-cool weapons are standard issue for alien grunts on an invasion mission. We almost certainly would be helpless, like fluffy little ducklings snapped up by a crocodile.

Think about it: any civilization that has mastered interstellar flight would be almost inconceivably advanced compared to us. Covering such immense distances on any reasonable timescale would require harnessing huge amounts of energy—energy that the invaders could dump right on our heads if they wanted to. Maybe they'd have matter-antimatter drives, like the Federation ships in *Star*

---

1   In a September 1987 speech to the United Nations, Reagan said, "I occasionally think how quickly our differences worldwide would vanish if we were facing an alien threat from outside this world."

Trek. Maybe they could warp space-time, making worm-holes at will. And here we are, still generating power by burning black rocks that we grub out of the ground.

So engaging an alien colony ship over Los Angeles or London, even with humanity's full complement of 15,000 nuclear weapons, would be comically fruitless, like fending off a charging rhino with a spatula.

We're probably flattering ourselves with the Christopher Columbus analogy here; the Aztecs, Incas, and Spaniards were all members of the same species, after all. (Despite their name, the Incas were a race of regular people, not octopus people.) An alien invasion of Earth may unfold more like the plunder of the Indian Ocean island of Mauritius by European seafarers and settlers, which quickly led to the extinction of the Mauritius owl, the broad-billed parrot, the Mauritian shelduck, the Mascarene gray parakeet, the small Mauritian flying fox, the Mauritian giant skink, Hoffstetter's worm snake, two types of giant tortoise, and more than a dozen other species, including the dodo.

"Even if you take into account classic military doctrine—like the defender always has an advantage, lines of communication, et cetera—it would be a no-contest, I'm sure," Pope said.

Classic military doctrine also calls for being prepared to tackle a range of conceivable threats, but Pope suspects that neither the US nor the United Kingdom has drawn up plans to deal specifically with an alien invasion. He said he never saw any such document—which he almost certainly would have helped draft or manage during his UFO days—in all his time at the Ministry of Defence.

Pope thinks not having a plan would be "a huge mistake," and not because it would make any real difference in battle—ET would still beat us like a rented mule, even if our flying-wedge pincer maneuvers were bullhorned to defcon perfection. (I'm not a military man, but I can talk the talk.) But a blueprint of sorts might help reduce the panic and chaos that an invasion would incite, thereby maintaining a functioning society for at least a little while longer. It could also keep post-apocalypse rat populations in check, by speeding the removal of corpses from the battlefield.

It's worth mentioning that the invaders wouldn't necessarily have to be big and toothy to do us in. Think about how the conquest of the New World went down. Europeans took over so quickly largely because native populations had no immunity against the diseases the colonizers brought with them, especially smallpox and measles. These germs killed huge numbers of indigenous people, causing their numbers to crash from perhaps 60 million throughout the Americas before contact to 5 million or so by 1650. (These numbers aren't well known, however; estimates of the Americas' pre-Columbian population range widely, from about 10 million to more than 100 million.)

Now, it's unlikely that random alien bugs riding an interstellar comet or asteroid could wreak such havoc. Infectious bacteria and viruses here on Earth have co-evolved with humans and other host species before making the jump to us. ET germs, on the other hand, would probably be so different from Earth life that they lack the molecular machinery needed to infiltrate our cells.

"They would just be there, like sand on the beach," Shostak said. "You wouldn't have to worry about them infecting your body."

But who's to say any alien bugs we may encounter would be random or naturally occurring? If aliens with designs on Earth knew enough about human biology, some scientists have warned, they could knock us out with a rain of genetically engineered microbes, then mosey down here at their leisure, plant a flag on the ruins of our civilization, and lounge on all the newly unclaimed mega-yachts until the aging, expanding sun burned our planet to a crisp.

SHOULD WE KEEP OUR MOUTHS SHUT?

Such worries about super-advanced, resource-hungry or just plain thoughtless aliens (like the Vogons in *The Hitchhiker's Guide to the Galaxy*, who blasted Earth to dust to make way for a hyperspace bypass) have sparked a fierce debate within the SETI community. Hawking and other like-minded folks have argued that we shouldn't beam signals out into the cosmos in an attempt to contact ET — or, at the very least, that we should be extremely careful about what we reveal in such messages, lest we expose our mega-yachts to alien lounging. Why betray our presence to potential enemies who could mash us into pink gravy if they wished? Even if the odds of such a signal reaching the ears of malevolent aliens were very low, we're talking about possibly inviting our own extinction here. So, active SETI (aka METI) is not worth the risk, the thinking goes.

"I don't think the dangers are high," Werthimer said. "But the problem is, [even] if it's a small risk, you have to multiply that by every human on the planet. If the risk is one in a thousand that, every time you transmit, ET is going to come eat us, then on average you're killing 7 million people. If the risk is one in a million, you're killing 7,000 people every time you transmit."

Astronomers have voiced other reasons for staying in listen-only mode, at least for now.

"We're just not grown-up enough to do this," Jill Tarter said. "If you start a transmitting program, you have to commit to it for thousands, or tens of thousands, of years to have it be successful. If you just transmit for five minutes, the signal's going to go right on by your intended target, and the chances that they're going to be looking at you in just the right way, at just the right time, are minuscule."

We may be able to tackle METI in a meaningful way after we've solved some of the problems that threaten to kneecap us as a civilization, such as runaway population growth and climate change, she added.

On the other side are people like Doug Vakoch. He acknowledges that keeping the METI effort going for many generations will be difficult, but he thinks that we should try anyway. Doing so could help our famously short-sighted species gain a bit of long-term perspective, which is a worthy end in itself, he said.

"If active transmissions go forward in a sustained way, if future generations are listening for a response thousands of years from now, whether they hear anything or not, it will have been a success," Vakoch said.

There are other possible reasons for humanity to take the initiative, he and others stress. For example, advanced aliens may be silently monitoring us from afar like cosmic peeping toms, waiting for us to show that we want to talk. Or they may currently view us as insufficiently interesting to trifle with—say, the way most of us regard ants (which is a shame, because ants are actually quite interesting). So our reaching out would be like an ant tugging on your sleeve and asking about your day. You'd want to talk to that little fellow, wouldn't you? Maybe even tell him or her something really cool and useful, like how the internal combustion engine and antibiotics work, or how to build an anteater trap? One of the hopes here is that messaging ET could not only spur the greatest scientific discovery of all time but also show that we're worthy of joining the Galactic Club. And a nice star-pattern blazer may not be the only perk of membership.

Advanced aliens "could help us resolve puzzles that we don't have the answers to," Avi Loeb said. "In principle, if they are a billion years older than we are in terms of technology, then we can get a shortcut for our advance into the next phase. We could save a billion years in our evolution."

The benefits of METI are potentially huge, folks like Vakoch and Loeb say. What about the costs—you know, that potential extinction-of-humanity thing? Couldn't the super-cool robot the aliens teach us to build be secretly programmed to kill us all? Well, advocates of active SETI tend not to be too worried about drawing evil aliens to our door. Any civilization close enough to reach out and whack us would already know we exist, having detected old episodes of *I Love Lucy* and all the other radiation we've been leaking into space for generations, Vakoch said. (Werthimer, however, counters that leakage is faint and much harder to detect than an intentionally beamed message.)

Vakoch also doesn't think ET would take any special malign interest in Earth or its resources, given how common rocky planets are across the cosmos. Shostak made this point as well. So proselytization and scientific curiosity are the only reasonable motivations for a visit, he said—either the aliens want to convert us to their religion, or they want to figure out what makes us tick.

"Neither of those is very compelling, given the cost of sending interstellar battle wagons to Earth," Shostak said.

Are you wondering about enslavement? Despite what you may have read, this doesn't seem to be a viable motivator, either. After all, a full-on spacefaring civilization would certainly be able to build robots that are better at mining unobtainium or pleasuring their lubricious sultan than we could ever be.

BE VERY AFRAID (OR DON'T)

How scared should you be of ET? That's impossible for me to say, because we have no idea what advanced aliens are actually like, if they even exist at all. It may just come down to whether you're an optimist, like Shostak, Vakoch, and Loeb, or a pessimist/risk calculator, like Werthimer and Hawking.

Though my feelings on this are far from settled, I probably lean more toward the optimists' camp. The "why would they bother with Earth?" rationale makes sense to me, as does some other informed speculation: all the headline-dominating bad news notwithstanding, humanity has been getting less violent and less aggressive over the centuries, as shown by Steven Pinker's 2011 book, *The Better Angels of Our Nature: Why Violence Has Declined*. If this is a trend that holds across civilizations throughout the galaxy—a big *if*, to be sure—then super-advanced aliens may actually be quite gentle. First contact may come in the form of a hug or a condescending pat on the head.

Besides, I'd prefer not to add another pressing existential threat to the already long list of things to worry about—climate change, nuclear war, the rise of artificial intelligence, nanotechnology running amok, widespread ecosystem collapse, and genetically engineered superviruses, to name a few. If one of these self-inflicted wounds takes us down in the near future, we won't get a pat on the head, and we won't deserve one, either.

Part II

Getting Out There

Chapter 12

Will We Colonize the Moon and Mars?

Good news for everyone who hates Earth: you could soon be able to forsake the planet of your birth, which has taken you for granted all these long years. All you have to do is become an astronaut, or perhaps a moon miner. Or get super rich.

In December 2017, President Donald Trump signed a directive instructing NASA to send people back to the lunar surface, which hasn't gotten a fresh bootprint since the Apollo 17 astronauts headed for home in 1972. (The stale bootprints are still there, though, since there's no wind to blow them away.) China is targeting the mid-2030s for its first crewed lunar landing, and European Space Agency officials have been talking for a while now about setting up an international "moon village"—a vision that very well may become reality in about the same time frame.

"I think that there are likely to be government-sponsored, international outposts on the moon's surface in the next 20 years or less," said space policy expert John Logsdon, a professor emeritus at George Washington University's Elliott School of International Affairs. He expects the United States to lead this lunar charge, because of the country's wealth and its space exploration leadership and heritage.

What will these moon pioneers do, you may ask? Lots of science stuff, like chiseling off chunks of rock and bringing them back to the lab for study. Peering up at the dazzling-bright stars through the moon's black, airless sky (and these stars wouldn't twinkle, no matter how lovingly you sang to them; stars twinkle in our night sky because of turbulence in Earth's atmosphere). Building radio telescopes to soak up far more black hole burps and pulsar pings than we could ever hope to hear on Earth—especially if the outpost is on the moon's far side, pointed blissfully away from all those terrestrial transmitters blasting *Law and Order* reruns and *The Fifth Element* into space. And, most importantly, learning to live off Earth, to prepare humanity for the pending jump to Mars. (More on that below.)

Any moon base will start out small, with just a handful

of (likely government) astronauts doing construction and research work. (Robots, the vast majority of them not evil, will help out a lot, of course.) But it may grow considerably: Rich folks may come visit for a spell, staying at the base or a nearby hotel—perhaps a big inflatable habitat built by Bigelow Aerospace—yep, the same company that may or may not be storing alien alloys for the US government—which aims to have one circling the moon by 2022, by the way.

Then there's that mining we talked about. The outpost will need dirt and rock, to build stuff (using 3D printers, mostly) and shield dwellings from temperature extremes and radiation. (The moon has no protective magnetic field or atmosphere.) So that omnipresent gray stuff will probably be the first target. But later, mining operations will target water ice, which appears to be plentiful at the bottoms of permanently shadowed craters near both lunar poles. And after that, mining robots could start grubbing up rare-earth metals, thorium (a radioactive element that serves as fuel in some types of nuclear reactors), and other materials that may even be valuable enough to ship back to the home planet for sale. If that business boomed, the robot workforce would expand, and more people would be needed to fix and maintain the machines. Fortune seekers from Earth could descend on the moon like 49ers sailing into San Francisco Bay back in the gold rush days.

Not all of these robot miners would be associated with the original lunar base, however big it may have gotten. A number of private companies, such as Moon Express and iSpace, plan to get in on the action as well with robotic spacecraft, starting with that water ice — and not just because it's good for drinking and bathing and growing space beans. The big idea is to split H_2O into its constituent hydrogen and oxygen, thereby making rocket fuel, which would be sold at off-Earth depots. Allowing spaceships to fill up their gas tanks on the go should help open up the heavens to exploration, the argument goes.

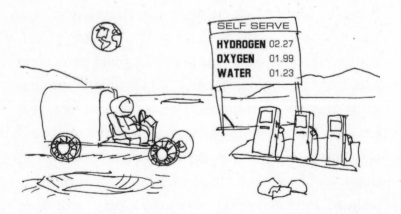

"There is no way to do large-scale activity in space without space resources," said Chris Lewicki, president and CEO of the asteroid mining company Planetary Resources, which also plans to start with water.

Sounds great! But space mining could lead to some space drama, if these companies end up making serious cash.

"We always fight over scarce, valuable resources," said Harvard astrophysicist Martin Elvis. "If there's enough money at stake, then the disputes could spread to Earth, and that would not be good. I'm hopeful that we'll avoid that, but I'm not confident."

Space resource disputes could linger or snowball, Elvis said, because there's no universally acknowledged authority to adjudicate them. The 1967 Outer Space Treaty forbids countries from making a territorial claim on the moon, asteroids, or any other celestial body, but it doesn't directly address mining and the rights of private companies. (People weren't thinking much about asteroid and moon mining during the Summer of Love.) The United States has taken this to mean that mining is A-OK, and has even passed legislation explicitly allowing American companies to profit from space resources.

But not everyone agrees. For example, some countries think we need an international licensing regime to keep tabs on space mining and to treat those operating outside of such a framework as cosmic scofflaws. A few nations could conceivably even pass laws prohibiting the sale of space resources within their own borders.

"They may consider it as similar to blood diamonds, or stolen art," said Frans von der Dunk, a professor of space law at the University of Nebraska–Lincoln. "Nobody knows where this is going."

Back to the moon outpost. How big will it get? One hundred people? Five hundred? A thousand? If a thousand, does it graduate from "outpost" to "settlement"? (People in the spaceflight community tend to prefer "settlement" to "colony." Colony implies a level of dependence on a larger ruling power back home, and it also evokes horrible mistreatment of native peoples.) I don't know the answers to these questions. But people will be back on the moon, and pretty soon, and that's good enough for me.

Actually—no, it isn't. I'm greedy. I want a settlement on Mars, too.

SETTLING THE RED PLANET

NASA aims to send a handful of astronauts to Mars in the 2030s to hunt for signs of ancient life and do other cool research and exploration work. But if we're talking Red Planet settlement, we need to talk about SpaceX.

Mars has long been in Elon Musk's sights. Back in the early 2000s, he hatched a plan to send a little greenhouse to the Red Planet, as a way to help get the United States and the world excited about Mars exploration. Green plants growing against a red-dirt backdrop, pioneering life on a cold and distant world would indeed make a great visual.

Musk was dead serious about this Mars Oasis project, even traveling to Russia to price rocket rides. But such research soon convinced him that the world needed cheaper and better rocket technology, and that he should develop that technology himself. So Musk, never shy, founded SpaceX in 2002, with the express goal of helping make humanity a multiplanet species, thereby lowering our chances of extinction. (If some enterprising dinosaurs had settled Mars long ago, after all, we might be eyeing dinosauroid canals on the Red Planet through our telescopes today.)

And now we know how he aims to do it. In the fall of 2017, Musk unveiled SpaceX's updated Mars settlement architecture—a huge, reusable rocket-spaceship duo called the BFR, which is short for "Big Falcon Rocket" or "Big F*cking Rocket." And I do mean huge: the BFR vehicle will be 348 feet tall by 30 feet wide, and the rocket will be the most powerful ever built—capable of launching more payload to space than even NASA's Saturn V moon rocket did back in the Apollo days, according to Musk.

SpaceX envisions the BFR as a do-everything system that will eventually replace the company's workhorse Falcon 9 and recently debuted Falcon Heavy rockets, as well as its Dragon space capsule. Phasing out this other gear will help make the BFR's development economically feasible, Musk has said. If all goes according to plan, BFR spaceships will soon be zooming back and forth from Earth to Mars[1] in huge waves, each one loaded up with about 100 people and all their Furbies and fidget spinners and whatever else Red Planet settlers are into.

How soon? Well, as of the time of this writing, SpaceX hoped to launch a cargo and scouting mission to the Red Planet in 2022, and then mount the first crewed mission in 2024. (This roughly two-year gap between launch opportunities is a result of orbital dynamics: Mars and Earth line up just right for interplanetary missions once every 26 months.) If everything works out, full-on operational passenger flights will follow

---

[1]   And to the moon. Musk has said the BFR system could ferry folks to a variety of destinations throughout the solar system.

before too much longer, taking more and more people, until we have ourselves a relatively self-sufficient, million-person city on the Red Planet, ideally by the end of the century.

SpaceX isn't drawing up blueprints for this city; it's just providing the transportation. Musk has compared the BFR's role in Mars settlement to that of the transcontinental railway, which spurred development throughout the American West in the 19th century.

Now, I don't want to give you the wrong idea: SpaceX isn't our only hope for getting off Earth in a meaningful way. For example, another billionaire-backed spaceflight company, Jeff Bezos's Blue Origin, also has a stated goal of helping get "millions of people living and working in space." Like SpaceX, Blue Origin is developing cost-slashing reusable rockets and has successfully landed and flown boosters multiple times. Blue Origin representatives have said that Mars is part of the company's long-term vision, but they haven't revealed many specifics about what exactly they intend to do.

MAKING A HABITABLE WORLD

The first Mars settlers will encounter a world that wants them dead and has at least four interesting and painful ways to make it happen: asphyxiation, freezing, radiation, and blood boiling (Mars's atmospheric pressure is so low that the gases dissolved in your blood would come bubbling out of solution right quick; in fact, this is what would kill you first, NASA planetary scientist Pascal Lee has said). So, at first, they'll probably spend most of their time underground.

But Marsonauts would doubtless eventually like to stroll around in the sunlight in shirt sleeves holding tennis rackets with sweaters knotted around their necks, just like we all do here on Earth. So they may try their hands at planetary engineering, terraforming their new home to make it more like their ancestral one, which only occasionally tried to kill them and their ancestors (via relatively blunt instruments such as bacteria, viruses, crumbly cliff edges, and leopards).

This would require, first of all, beefing up Mars's now-wispy atmosphere with lots of heat-trapping greenhouse gases—stuff like CO_2 (already the dominant component of Red Planet air), water vapor, ammonia, and fluorocarbons.

Over the years, settlement enthusiasts have floated many possible ways to do this. You could nudge water-rich comets or asteroids onto collision courses with Mars, for example, or detonate a bunch of nukes above the planet's poles, vaporizing the H_2O and CO_2 locked up there now as ice. Or, if you're not so into violence, you could bake off that polar ice using giant orbiting mirrors, or spread dark dirt atop the region to soak up sunlight more naturally.

The settlers could also build factories that pump super-potent fluorocarbons into the air—basically, intentional industrial pollution. (This would also be a sign of intelligent life that could be detected by distant aliens; see chapter 5.) Or they could rely on bio-engineered microbes or fancy self-replicating nanomachines to do this job, depending on how advanced they are, said Mars Society president Robert Zubrin, a long-time and vocal advocate for getting humans to the Red Planet, and author of the influential 1996 book *The Case for Mars* (which includes a chapter on possible terraforming strategies).

Even if this warming work does the trick, however, you're still only halfway done: You need to seed the red dirt with lots of plants and/or photosynthetic microbes to get some oxygen flowing. And even when you're done, you're not done. Mars has no global magnetic field, so the solar wind will tear at the newly thickened air constantly. The settlers will therefore have to perform some maintenance work from time to time to keep the planet livable.

Could all of this work? Zubrin thinks so. And he's

optimistic about the time frame, too, especially considering how technologically capable our near-future Mars explorers are likely to be (they of the postulated bioengineered microbes and self-replicating nanomachines). "I think that Mars will be terraformed within a couple of hundred years, and, when it's done, it will be done by techniques that we consider science fiction today," Zubrin said.

If you're a little uneasy with the thought of rejiggering another planet to meet our own needs, you're not alone. It makes me squeamish, too. If Mars microbes are still clinging to life on the frigid world, we could fundamentally alter their evolutionary path, or perhaps even wipe out an entire ecosystem with all our tinkering.

Zubrin, however, is unconcerned. He thinks that native Martians, if any do indeed still exist, will be just fine in their moist subterranean redoubts, no matter how baroque our additions to the surface may be. And expanding Earth's biosphere to Mars, allowing the profusion of all sorts of new and exotic life-forms, would actually be a good thing, he said. "A whole new living world created—to not do that would be an abdication of responsibility to the biosphere, to life."

NOT SO FAST?

There are some serious obstacles to making this grand vision of a million or more people blissfully going about their business on Mars, playing tennis or pretending to enjoy doing so, a reality. One is the high cost of launch, which companies like SpaceX and Blue Origin are working to slash with their big, reusable rockets (Blue Origin is scheduled to debut its New Glenn heavy lifter in 2020, and it's also working on an even more powerful rocket called New Armstrong). But another is the whole living-off-Earth thing. We've never done it before—astronauts' relatively short stints aboard the International Space Station notwithstanding—so we don't know how hard it will end up being.

"We have never done any studies that look at how much gravity is enough gravity," said Kris Lehnhardt, senior faculty at the Baylor College of Medicine's Center for Space Medicine. "So, the moon is one-sixth [that of Earth]; Mars is three-eighths. We don't know where that tipping point is where you have enough gravity to maintain your normal body systems."

We don't even know if women can get pregnant in space, or if a fetus can develop normally away from our home planet, he said. That could be a big deal, obviously:

no Mars babies means no Mars settlement, at least not a self-sustaining one, anyway.

And some folks think terraforming Mars will be a lot more difficult than Zubrin imagines. For example, Ruvkun is dubious that we'll be able to tune the dial just right using microbes, bioengineered or not.

"Good luck having it work so that it's perfect for us," Ruvkun said. "You're basically replaying evolution. And if you replay evolution, are you always going to develop photosynthesis? Maybe not; you might have a hyperthermophilic ecosystem. The bacteria that invented photosynthesis almost poisoned the Earth. They had 99 percent extinction; they totally changed the atmosphere. Everything died but them and their descendants—us."

So, Ruvkun is definitely bearish on Mars settlement. (He did say, however, that Musk could conceivably make money turning Mars into a graveyard: "If you could bury people on Mars for $1,000 a pop, you could beat what it costs to bury people here.") And Logsdon is skeptical of Musk's plan as well. "There's no reason to have a million people living on Mars," he said. "Elon starts with the assumption that having a million people on Mars is a good thing, but he's never offered a rationale for it. And neither has anyone else."

Again, I have no crystal ball. I don't know who will end up being right, the skeptics or the optimists. But my money's with the optimists, though I suspect it'll take longer than they hope. That's the way things usually work out, anyway.

LIVING IN A MOMENT

I want to end this chapter with a little reminder: in case you couldn't tell from all the stuff we just talked about, we live in exciting times. Enjoy it. Appreciate it.

You've seen videos of SpaceX's Falcon 9 rocket landings, right? A rocket that launches a satellite to space, and then comes back down for a vertical landing? On a robotic ship at sea? (Yes, sometimes they come back on terra firma, but those are marginally less spectacular.) And Falcon Heavy's maiden launch in February 2018, which put Musk's red Tesla Roadster, driven by a mannequin dubbed Starman, on an orbit that goes out beyond Mars?

Blue Origin has a bunch of rocket landings under its belt, too, of its New Shepard suborbital vehicle. And the New Shepard ride-along test dummy has an even cooler name than Starman: Mannequin Skywalker.

New Shepard may start flying customers on suborbital jaunts any day now, and Virgin Galactic's SpaceShipTwo is almost ready to go as well. An Arizona company called

World View is developing a system that will take tourists on smooth balloon rides through the stratosphere. And SpaceX and Boeing could start ferrying NASA astronauts to and from the International Space Station as early as next year.

Also, both SpaceX and Virgin Galactic have said they aim to get into the "point-to-point transportation" game. Which means that, before too much longer, you could end up taking a spaceship on your next flight from LA to New York—and get there in less than half an hour.

Planetary Resources and another company, Deep Space Industries, are developing gear to do real-life space mining. The miniaturization of electronics is allowing the production of very small, capable, and cheap satellites, opening up space to huge amounts of research and exploration. The San Francisco company Planet has dozens of sharp-eyed satellites eyeing the Earth, for example, and each one is the size of a loaf of bread.

And did you know that you're living in the age of off-Earth manufacturing? A company called Made In Space has launched two 3D printers to the space station, as well as a machine designed to make a super-valuable optical fiber in orbit. If a test run goes well, Made In Space plans to churn out bundles and bundles of this high-performance fiber in space, haul it back down to Earth, and sell it. For lots of money.

Being able to do this sort of thing on a large scale would be a very big deal, of course. If we can build spaceships and fuel depots and solar arrays and habitats and wheely shoes in space, using space resources, we can cut the ties with Earth.

"It's literally the difference between going on a camping trip in the woods and going and settling," said Made In Space CEO Andrew Rush. "It's the tools that you take with you."

I could go on and on and on. The point is that things are happening now. Big things. Private spaceflight is coming into its own, and this new surge may finally, after decades of false starts and false hope, get us out there for good.

"I do think that we're living through, in my opinion, sort of the second big moment in space exploration, or space development," said Virgin Galactic CEO George Whitesides. (The first big moment, he said, was the first decade of human spaceflight, from Yuri Gagarin through the Apollo moon missions.) "We're still on the front end of that. I think that this push could be another 100 years."

Chapter 13

Can We Go Interstellar?

Maybe you're even greedier than I am, and you don't want to stop at Mars. You want to get people all the way to Proxima b, 4.2 light-years away, and you want to get them there *fast*—in less than 4.2 years, if possible.

Surprisingly, that is possible—at least in theory. Sure, nothing can travel faster than light through space; that would violate Einstein's law of special relativity. You know about special relativity, because you've seen its most famous equation: $E = mc^2$. This tells us that mass and energy are equivalent, and related via the speed of light (c, which is about 186,000 miles per second). Special relativity also tells us, among many other things, that an object's mass increases as it speeds up, and that even the tiniest object in the universe would become infinitely massive at light speed. This is impossible, because you'd need infinite energy to accelerate an infinite mass. And infinite energy is not a thing, even in our indescribably weird universe. (Light can travel at light speed, by the way, because photons are massless particles.)

But we might be able to cheat—to manipulate space-time in such a way that we could zip somewhere faster than even a light beam could. I'm talking, of course, about the fabled warp drive you know so well from science fiction.

In 1994, Mexican physicist Miguel Alcubierre gave the idea a solid mathematical backbone, showing that a starship could warp by contracting the space-time in front of it and expanding the space-time behind. (No joke: the basic idea came to him while he was watching *Star Trek*.) This would create a sort of wave that the ship could ride at speeds exceeding that of light, like the most badass surfer in the history of the world. This doesn't break any cosmic rules; as we discussed in chapter 3, nothing prohibits space-time itself from expanding faster than the speed of light. Indeed, physicists think this happened during the period of inflation right after the Big Bang, when the universe ballooned from incomprehensibly microscopic to gigantic in a few tiny fractions of a second.

Sounds great, right? But there are a few problems.

First of all, to create this gnarly space-time wave, you probably need lots and lots of negative energy, and not the kind you soak up against your will from your jealous and petty "friends," but the super-strange physics kind. Now, this crazy stuff does seem to exist. In the lab, researchers have shown that quantum effects create areas of negative energy density. Specifically, if you put two metal plates extremely close together, they move toward each other ever so slightly. This Casimir effect results from the fact that fewer virtual particles pop into existence between the plates than beyond them, so the net particle push is inward. (The quantum world is freaky. Don't try too hard to understand it; it can break your brain.)

Will this phenomenon ever rise beyond the status of gee-whiz novelty? Who knows. "We have no idea if we can harness it," Alcubierre said.

Dark energy—the mysterious stuff that's driving the universe's accelerating expansion—could also be useful to warp drive engineers, if it turns out that it's some kind of field that can be manipulated. But we don't know that it is. In fact, the simplest explanation, Alcubierre said, is that dark energy is a constant, something that's just sort of there in the background, like extras in a movie wedding scene. (In fact, dark energy may be Einstein's famous "cosmological constant." The great physicist introduced this idea in 1917 to help explain why the universe is static, which was the prevailing view at the time. When it became clear that the universe is actually expanding, Einstein declared the constant his biggest blunder. But he may have been on to something after all.)

But there's an even bigger issue with warp drives: the horizon problem. The starship would be sitting in a sort of bubble, with space-time moving faster than the speed of light all around it. The ship would therefore be completely cut off from the region in front of it—meaning there would be no way for it, or its captain, to set off the negative energy, or ignite or initiate or detonate it, whatever the proper verb is for this oddball entity. So the warping starship would have to follow a trail of negative-energy breadcrumbs laid out by another spacecraft that came before it. Which kind of defeats the purpose of super-fast travel.

"My gut tells me that, if it's possible at all, it's going to be extremely impractical, and not for a very long time," Alcubierre said of warp drives. "I would put more of my money on wormholes."

Ah, wormholes! Another sci-fi staple, and another way to cheat. Wormholes are cosmic shortcuts, tunnels joining two different spots in space. Think of a folded piece of paper, through which a hole is poked with a pencil.

It's much quicker for an ant to duck through the hole, obviously, than to trek all the way around on the surface of the sheet.

But there are some serious issues with wormholes as well. First of all, nobody knows for sure that they actually exist. And future wormhole builders/manipulators would have to conquer one of the problems that also bedevils would-be warp-drive engineers: negative energy. You'd need negative energy to keep a wormhole open for any reasonable amount of time; otherwise, it'll snap shut on would-be starshippers like a cartoon giant clam.

So, I wouldn't count on a wormhole trek to Proxima b anytime soon. Indeed, Alcubierre's endorsement of these space-time oddities was qualified: "Again, we don't know how to make a hole in space," he said.

But we don't have to go superluminal to spread out among the stars; we could just go really, really fast. And that's a lot more achievable in the not-too-distant future. Alcubierre cited nuclear-fusion rockets, saying they could very well accelerate a starship to 10 or 20 percent the speed of light. That would make the trip to Proxima b quite manageable—just 40 to 80 years, instead of the 75,000 years it'd take with traditional chemical propulsion. Now, we don't have fusion rockets yet. Indeed, we haven't even built power-generating nuclear-fusion reactors, despite trying to do so for decades. (Current nuclear power plants are built around splitting rather than fusing atoms.) But we have built nuclear-fusion bombs, so that's something.

Nuclear-propelled spaceships are not a new idea. In the 1950s, for example, the US Defense Advanced Research Projects Agency (DARPA) began working on Project Orion, which seems like something out of a Cold War fever dream. The plan called for getting astronauts to Mars just two weeks after launch by detonating a series of atomic bombs behind a spacecraft. DARPA abandoned Project Orion in the mid-1960s, shortly after the Partial Test Ban Treaty made it illegal to explode experimental nukes in space. But the idea didn't die.

In 1968, Freeman Dyson, who had worked on Orion, published a paper positing that a souped-up version of the project could be used for interstellar flight. Scientists kept developing interstellar nuke concepts over the next few decades, working on Project Daedalus in the 1970s and Project Longshot in the late 1980s, but no such ships have ever come close to being built.

That's not to imply that nukes have never made it to space in any form, however. Indeed, many deep-space robotic explorers, including NASA's Mars rover Curiosity and New Horizons Pluto probe, power their scientific instruments with radioisotope thermoelectric generators, which produce electricity using the heat emitted by the radioactive decay of plutonium.

You may be wondering about antimatter. And you should! It is wondrous stuff—the super-rare, bizarro version of the normal matter we're familiar with. Whereas a proton has a positive charge, for example, an antiproton has a negative charge. Matter and antimatter annihilate each other when they meet, in a reaction that's 100 percent efficient. That is, every last scintilla of matter gets converted to energy (specifically, gamma radiation). This is an amazing fact, so we're going to dwell on it for a second. For context, the fusion reactions in the core of the sun, which everyone always oohs and aahs over, are just 0.7 percent efficient.

So you can see why matter-antimatter reactions get people so excited (and why the creators of *Star Trek* invoked them as the power source for the USS *Enterprise*'s warp drive). But, again, there are big problems. There's very little antimatter out there in nature, and the tiny bits that do exist don't last long; they're always running into the far more abundant regular matter and blowing up. So the stuff would have to be specially created for an interstellar mission. Physicists have made antimatter using particle accelerators here on Earth, but only enough to fill a mouse's snout; creating enough to propel a spaceship would probably cost hundreds of billions of dollars, if not more. And you'd have to build a pretty fancy gas tank, because if a speck of antimatter brushed the side of a normal container, the whole ship would go kaboom.

TO BOLDLY GO
WHERE NO BOT
HAS GONE BEFORE

GOING ROBOTIC

We don't have the propulsion thing ironed out yet for an interstellar settlement mission. And there are other hurdles we need to clear as well—for example, figuring out how to keep people alive, healthy, and relatively happy (or at least not murderously disaffected) for years at a time in deep space.

Scientists and engineers are trying to solve these life support and maintenance issues, testing out ideas on the International Space Station and in labs here on Earth. But fully vetted solutions are still a ways off.

"If you talk about the prospects of colony ships, whether they're just in orbit or going anywhere—rough-order guess, it might be half a century before we even have that possible," said Marc Millis, the former head of NASA's Breakthrough Propulsion Physics Project and founder of the nonprofit Tau Zero Foundation, which looks into ways to make interstellar spaceflight happen.

(In case you were wondering: one longed-for colony ship feature—full-on suspended animation, as seen in movies like *Aliens* and *Passengers*—remains a sci-fi dream for now. But inducing a temporary hibernation-like torpid state in astronauts is very much in play in the not-too-distant future. By the end of the 2030s, it may well be feasible to put space travelers under for several weeks at a time by lowering their body temperatures, said John Bradford of SpaceWorks Enterprises, who leads a group that has received several rounds of NASA funding to research this strategy. Cyclically inducing torpor would lower the psychological burden on voyaging crewmates and drastically reduce the amount of stuff needed to support them along the way, among other benefits, Bradford said. Crewed interstellar missions would still require a propulsion breakthrough, he added, but "journeys within the solar system are all on the table with this technology.")

It would be much easier, then, to just go robotic. And we may be on the doorstep of doing just that.

The $100 million Breakthrough Starshot project—another ambitious program bankrolled by Yuri Milner—is developing a "directed energy" system that would blast tiny sail-equipped spacecraft to about 20 percent the speed of light using powerful Earth-based lasers. And I do mean tiny: the probes' bodies will be about the size of postage stamps, though their unfurled sails will be about 13 feet on a side. These sails harness the pressure exerted by photons, much as seagoing ships' sails harness the wind. (Yes, photons impart momentum despite having less mass than a lapsed Catholic.)

The basic idea behind space sailing has been around since 1964, when sci-fi legend Arthur C. Clarke proposed it in a short story. But the technology is no mere flight of fancy: Multiple spacecraft have been pushed along (albeit relatively slowly) by sunlight pressure, beginning with Japan's Ikaros probe in 2010.

The Starshot team aims to launch its first probes toward Proxima b in the next 30 years or so. The long-term goal calls for sending thousands and thousands of these nanocraft to a variety of nearby planetary systems, to look for signs of life, take photos, and do a variety of other cool interstellar exploration work. It'll be tough to pull it off; challenges include keeping the sailcraft steady enough to catch its laser ride all the way out into deep, deep space and fitting onto the tiny probe a transmitter powerful enough to relay all the gathered data back to Earth. But it's possible, and on a reasonable time frame. We may all see up-close photos of Proxima b about 50 years from now, at about the same time the Chicago Cubs win their next World Series title. That is incredibly cool.

So cool that I'm going to say it again: Interstellar exploration! (Technically, we've already checked off this box. NASA's Voyager 1 probe popped free into interstellar space in August 2012, and its twin, Voyager 2, will do so soon as well. But these probes, both of which launched in 1977, won't reach another star system for tens of thousands of years.)

Are you rolling your eyes? Well, keep in mind that we're just 60 years into the space age. If we do manage to avoid offing ourselves, our exploration future could be Sirius-bright.

"All throughout history, people have been saying things like, 'Heavier-than-air flight is impossible,' 'We'll never get to space,' 'We'll never send humans to the moon,'" said Richard Obousy, director of Icarus Interstellar, a nonprofit that aims to make interstellar flight a reality by the year 2100. "And then we come up with some profound breakthrough that nobody anticipated, and all of a sudden, the impossible becomes obvious."

WHAT DO WE WANT?

We should take a step back here and articulate what exactly it is we hope to achieve with these exotic deep-space treks, because our goals should color our strategies.

For example, if the chief drivers are the desire to put a human imprint on another world and our innate needs to see over the next hill and stick it to everybody in high school who thought we'd never amount to anything, then maybe we should put lots of cash into nuclear fusion research. But if the chief aim is scientific curiosity, then little robotic probes like Breakthrough Starshot's cosmic sailors provide the most bang for the buck.

And if the main goal is mere survival, escape from our unfaithful bastard of a sun, then we should pour more resources into life support tech. If we can build huge colony ships with turnip-filled greenhouses and basketball courts and space toilets that are actually comfortable, after all, we don't need a planet to roam around on. We can just live in free space, drifting forever through the dark emptiness like anglerfish in the deep.

Millis made this basic planning point and stressed that, because of the uncertainty surrounding our goals for interstellar flight and the feasibility of the various technologies that could make it happen, we shouldn't try to a "pick a winner" right now. For example, let's not pour every penny into fusion rockets or laser sails; let's keep working on all of the promising tech and see what happens.

Millis also raised another issue, one that's rather depressing.

"Unfortunately, the greatest challenge to interstellar flight isn't about the technologies; it's about societal maturity," he said. "If you look at the energy levels, even for something as small as Starshot, you're talking destruction-of-humanity type of energy levels. I mean, if that thing was weaponized, it would be very bad."

When will we be ready to handle the technology that could take us out among the stars? Your guess is as good as mine. But I'm not terribly optimistic. After all, we haven't exactly covered ourselves in glory with our management of fossil fuels, which are a baby-step technology by comparison.

Speaking of survival: our current notions may be moot, and quaint, by the time we're ready to leave our solar system in earnest. For example, we may have already passed through the singularity and merged with artificial intelligence by that point. This would both greatly accelerate our technological development and obviate the need for ships with space toilets and basketball courts. And we wouldn't really care how fast our ships could go, since we'd be effectively immortal at that point (as long as we had access to energy and raw material, so we could repair and replicate the machines that constituted us).

But say we're worried about saving our original meatsack selves, and we just can't get the propulsion tech right, or the life support, and the space toilets keep backing up. We still have options. For example, we could engineer bacteria to carry our DNA, and then launch them out among the stars. Reassembling that "library" of genes into a functional human seems crazy and impossible now, but maybe it'll be a breeze in 10,000 years. Maybe the engineered bacteria will do it themselves when they flutter down on a properly pleasant planet. Maybe we'd have to launch a machine along with them. Maybe we'd count on Ned Flanders-friendly aliens to do it for us. Who knows?

This sort of deep-future prognostication is pretty futile; even the most talented visionaries fail spectacularly when they look too far out. Jules Verne, for example, couldn't predict the power and utility of rockets; his lunar explorers in the 1865 novel *From Earth to the Moon* were shot out of a giant cannon.

And if humanity's impending disappearance from the cosmic scene keeps you up at night, well, we can actually do something to soothe your jangled nerves right now: encode all of our DNA info in radio waves and beam it out into space, however much this may anger the folks who worry about malevolent aliens.

"Do you actually go extinct if your genome sequence has been sent off into the cosmos?" Ruvkun said. "Somebody could resynthesize us if they could figure out how to reprogram a cell with us. Are you really extinct if that's the case?"

Chapter 14

Will There Be a *Homo spaciens*?

Going to Mars will change us—and not just in a touchy-feely, metaphysical way.

Over time, we should expect a fair bit of evolutionary divergence between Mars settlers and the human population on Earth, according to Rice University biologist Scott Solomon, who examined this possibility in his 2016 book *Future Humans: Inside the Science of Our Continuing Evolution.* That divergence will start unspooling at the outset, thanks to something called the founder effect. No matter how or when it happens, the Red Planet will be settled by a relatively small group of people who are not perfectly representative of the entire human population. For example, it's a pretty safe bet that Mars pioneers will be atypically adventurous and risk-tolerant, so Muskton (the odds are decent that the first Red Planet burg will be named after SpaceX's CEO) will likely feature more rock-climbing gyms and bordellos per capita than cities here on Earth.

And those initial differences will snowball, because Mars and Earth are very different worlds. The Red Planet is much smaller, and the force of gravity on its surface is therefore just 38 percent of the pull we feel here on Earth. Mars also lacks a global magnetic field, a thick atmosphere (though we could remedy that to an extent via terraforming), and a protective ozone layer, so it gets hammered a lot harder than we do by space radiation—UV light and charged particles from the sun, as well as super-energetic cosmic rays zooming in from outside the solar system.

This damaging radiation could cause higher mutation rates in the DNA of Mars settlers, Solomon said. Mutations increase genetic variability, so evolution may proceed faster on the Red Planet than it does here on Earth. What sorts of changes could we see over there? Well, for one thing, natural selection might adjust skin tone on the Red Planet, to help settlers cope with that serious radiation load. (Even if they live in modified caves or lava tubes, as seems likely, the pioneers will still have to spend some time on the surface—to tend to their crops and attend spring equinox chili cook-offs, for example.) This may lead to dark skin via increased production of melanin, just as we see among some peoples here on Earth. But other pigments could potentially be pressed into service as well, including carotenoids, the molecules that give real carrots—as opposed to those purple artisanal weirdos—their color, according to Solomon.

Mars settlers may also eventually sport thicker bones than their ancestors, Solomon said. That's because, as research on astronauts in low-Earth orbit has shown, bones become less dense and more brittle in low-gravity conditions. So Red Planet pioneers with abnormally stout skeletons may do abnormally well, throwing down monster dunk after monster dunk in games of Marsketball while their owned opponents roll around in the dirt clutching their broken femurs and moaning. And Marsketball would be awesome, by the way. If you kept the hoop at the standard 10 feet, all but the most sessile among us could dunk, since you can jump about 2.5 times higher on Mars than you can here on Earth.

And Martian colonists may not be plagued by plagues. The long cruise to the Red Planet could serve as a quarantine, keeping nasty germs from getting a foothold in the settlement, according to Solomon.[1] Muskton probably wouldn't have to worry about the next Ebola or West Nile emerging from the Martian wilds, which appear to be free of viruses and bacteria, let alone any chimps, birds, mosquitoes, or bats to incubate or transmit them. So, if the settlers left their mammal friends at home—the ones we like to eat, as well as the ones whose bellies we like to rub and ears we like to tousle—they could conceivably banish infectious disease to the memory hole. (The pioneers could go vegan, or eat bugs rather than cows and pigs. Insects are much further removed from us evolutionarily and therefore less likely to pass pathogens on to us.) The settlers' immune systems

---

1 Our microbiome—the collection of microbes that swarm all over and inside us—could also suffer greatly, since we pick up most of these little guys from the environment. A healthy microbiome is key to a healthy person, so this development could be a bad one for Mars settlers.

might then wither like a snipped umbilical cord, eventually atrophying into vestigiality. White blood cells could be the new tailbones.

"If that were to happen, if, somehow, a disease were to be introduced to Mars, it would be completely devastating," Solomon said. "That would set up a situation where any contact between Earth and Mars would be extremely dangerous. Steps might have to be taken to basically eliminate any chance of having contact. Even if there are shipments going back and forth, even if there are people going from Earth to Mars, perhaps they don't ever come into contact with one another."

This scenario would lead to a cessation of gene flow between Earth humans and Mars humans, and speciation could soon follow. How soon?

"I hate to ever put numbers on it, because it's still such a speculative scenario," Solomon said. "But you'd be talking about at least several hundred to, possibly, several thousand generations."

This putative outcome doesn't seem to jibe with our experiences here on Earth, where small bands of pioneers have repeatedly settled new lands without ever diverging into new species of hominid. For example, Native Americans and aboriginal Australians remain in the *Homo sapiens* fold despite having lived in relative isolation on their newfound continents for about 15,000 and

50,000 years, respectively. But we can take this comparison only so far: North America and Australia are still part of familiar old Earth, so the environment wasn't pushing those long-ago explorers to diverge nearly as powerfully as harsh, weird Mars will.

Solomon cautioned that nobody can predict how evolution will proceed in the future. Indeed, some folks have a different take on our relationship with those future inhabitants of Muskton. For example, Mars Society president Robert Zubrin thinks the settlers will develop one or more unique Martian cultures but will not radiate into a new species; they'll just be too close to Earth, with too much contact. He does think this will happen with interstellar settlers, however—partly because of the inevitable cultural differences that will arise.

"We're going to have the power, in principle, to control our evolution, to genetically engineer and influence our children," Zubrin said. "If we have established ourselves in new star systems, in some places, people will probably say, 'That's a great idea; let's do that.' In others, they'll say, 'Oh, that's immoral. We should not do that.' So, whether they do it or don't do it, it's going to cause divergence."

Such divergence, he said, could lead to a *Star Trek*–like panoply of humanoids that differ from each other in just a few trifling respects, such as the color and scaliness of their skin or the number and size of the bumpy ridges on their foreheads. You know, whatever look becomes fashionable on those deep-space outposts, so far from the dominant original culture and its homogenization machine. Hopefully hipster skinny jeans won't make it all the way out to GJ 273b.

Of course, the lack of gene mixing among colonists and their forebears on Earth would be an even bigger factor in our species' interstellar radiation—if there are still any genes around to be mixed. We may have advanced to cyborg/sublimated consciousness form by the time we start moving out among the stars.

Chapter 15

Is Time Travel Possible?

All of us have fantasized about traveling back to mid-1970s Munich, finding the house where young Lou Bega lives, and replacing any musical instruments inside with soccer balls and chemistry sets, thereby creating an idyllic future in which we never have to hear the squish-thumping opening strains of "Mambo No. 5" ever again. But is this possible?

Maybe. Sort of. Perhaps?

I'll try to explain. We need to start with Einstein's 1915 theory of general relativity. And it's a beautiful theory, describing gravity in a way that can be explained with bowling balls and trampolines. Think of space-time as a flexible sheet, like a trampoline, and think of all massive objects, such as planets and stars, as bowling balls. When you place a bowling ball on a trampoline, as happens at the best parties, the trampoline droops. If you roll a marble onto it, the marble spirals toward the bowling ball. This is gravity. The theoretical physicist John Wheeler stated the basics of general relativity far better than I ever could in his 1998 book *Geons, Black Holes, and Quantum Foam: A Life in Physics*: "Space-time tells matter how to move; matter tells space-time how to curve."

As you might imagine, general relativity is a bit more complicated than I've let on so far: It consists of lots of equations. There are many different ways to solve these equations, and these various solutions lead to many types of space-time geometries, some of which have pretty weird and interesting consequences.

For example, in 1949, the Austrian American mathematician Kurt Gödel found a solution to Einstein's equations that contained "closed timelike curves"—basically, cosmic loops that allow travel backward through time. (Remember, general relativity says space-time can be warped and distorted.) But these particular loops could exist only if our universe were rotating, which we now know that it is not.

And in 1974, the American physicist Frank Tipler determined that a general-relativity solution devised decades earlier also permitted time travel—provided the space-time in question harbors an infinitely long, rotating cylinder, which could drag you back temporally if you zoomed around it. Unfortunately, our sharpest-eyed telescopes have yet to turn up any signs of infinitely long cylinders, rotating or otherwise. And building one would be, shall we say, challenging.

But we could lower our sights a bit and just use wormholes.[1] Special relativity tells us that time slows down the faster you move through space, and this time dilation effect is pretty dramatic when you get up near light speed. For example, let's say you take a sightseeing trip to the TRAPPIST-1 system, aboard a starship that goes 99.9 percent the speed of light. It'll take you roughly 80 years to get there and back, from the perspective of the poor saps on Earth who didn't get to go. But for you, only about 3.6 years will pass. You'll come home with lots of great photos and an exotic red-dwarf tan, and we'll all be dead.

What does this have to do with wormholes? Well, you could theoretically imprint a time difference on a wormhole's two mouths, by flying one of them around really fast for a while. A brave traveler could then hop into the future by zipping through the wormhole in

---

1   I'm going to ignore faster-than-light travel, which could also enable time travel, because it's prohibited by special relativity (at least for massive objects like you and me).

one direction, and journey into the past by taking the other route.

The bad thing about this kind of time machine—I know, first-world problems—is that it could never take you back to a point before its invention. So, if you first popped such a wormhole open in 1986, the best you could ever do is miss the premiere of *Rocky IV* by a year. This is a possible retort, by the way, to folks who cite the lack of temporal tourists as evidence that traveling backward in time is impossible: maybe they can't get to our era because we haven't invented a time machine yet. (And there are other possible answers, of course. For example, maybe time tourists *are* here and we just don't know it. Or maybe our future selves are just wise enough not to do crazy stuff like mess with our timelines.)

I think it's time to talk about the grandfather paradox. It's a staple of all time travel discourse, so you've probably heard it before: you travel back in time and kill your grandfather before he had a chance to seduce your grandmother with those Andrews Sisters tickets. How is this possible? The murder would prevent your own existence.

Such issues potentially decouple cause and effect, making our universe fundamentally absurd. So, many scientists regard time travel as a sort of novelty— something that's mathematically possible in Einstein's equations but irrelevant to our existence here in the real world. And this isn't just a logic-based argument: "In all the examples that general relativity is able to provide, there are other physics that take over and prevent that from happening," said Paul Sutter, an astrophysicist at Ohio State University.

Stephen Hawking even proposed a "chronology protection conjecture," which basically postulated that the laws of physics will conspire to prevent time travel on all but the smallest spatial scales. For example, weird quantum effects may destroy your wormhole before you can use it.

"I think most physicists today will tell you that it's impossible to travel to the past—it doesn't make sense," physicist Miguel Alcubierre said.

Most, but not all. For instance, some folks cite two possible time travel paradox escape hatches. The first is the "banana peel" explanation, articulated by Allen Everett and Thomas Roman in their 2011 book *Time Travel and Warp Drives*: every time you go to stab your sleeping grandfather in the neck, you slip on a banana peel or get bitten by a raccoon or something. You just can't do it; something will always prevent this fundamentally unnatural act from occurring. And the second invokes the many-worlds interpretation we talked about in chapter 3: you can indeed kill your own grandfather in the past—but in a different timeline than the one that leads to you (or at least the "you" that you regard as you).

So the field is still pretty open, and getting definitive answers—knowing if the chronology protection conjecture is on the money, for example, or if you really could use spinning black holes as time machines, as some physicists have suggested—likely requires marrying gravitational effects on the macro and super-micro scales.

"We do need a theory of quantum gravity," Sutter said.

Time travel into the past is sexy and all—if it's possible, you could go back and club a dinosaur, or whup Julius Caesar in Trivial Pursuit. But what if I told you that you can time travel into the future, and that you're doing so right now? Because you are! You're marching into the future at the constant and boring rate of one second per second.

OK, that was a cheap—and quite lame—trick. But time travel into the future *is* a thing, thanks to time dilation. And astronauts experience it all the time, though on very subtle scales.

Take NASA astronaut Scott Kelly and cosmonaut Mikhail Kornienko. The duo stayed aboard the International Space Station from March 2015 to March 2016, on a mission to help researchers learn more about the physiological and psychological effects of long-term spaceflight so they can better plan for crewed journeys to distant destinations such as Mars. During that stretch, the spacefliers were zipping around Earth at about 17,500 miles per hour, seeing one sunrise every 90 minutes or so.

Now, Scott Kelly has an identical twin brother, Mark (who is a former NASA astronaut himself); indeed, that's part of the reason Scott was chosen for the 340-day mission. Mark stayed on Earth during Scott's space jaunt, serving as a control against which to gauge any changes spaceflight induced in his brother. Because Scott was moving so fast for so long, he aged a bit less than Mark did, at least by Earth standards. Mark started the mission 6 minutes older than Scott; by the end, the elder twin had tacked on an additional 5 milliseconds to his seniority.

Time dilation also occurs in strong gravitational fields—an effect shown in the 2014 film *Interstellar*, when the colonization scout team investigated a potentially habitable planet near a supermassive black hole. (For every hour the explorers experienced on that exotic world, 7 years passed on Earth, according to the film.) So if Scott Kelly wants to become a much younger brother, he could lobby for a seat on the next crewed black hole mission coming down the pike, or at least a lengthy stay near Jupiter's cloud tops.

Chapter 16

What Will Happen to Us?

I'd like to end on a light and cheery note: a brief discussion of our impending doom. For we are indeed doomed. Everyone and everything in the universe is going to die, eventually. Let's see how!

Remember from chapter 6 how, billions of years ago, an increasingly bright and powerful sun turned Venus from a Maldivian vacation world into a scorching hellscape? Well, that's going to happen to Earth, too. The sun is still brightening, and, about 1 billion years from now, it'll be strong enough to start boiling off our oceans. A runaway greenhouse effect will start rolling in disturbingly short order, and Earth will dry out and become hot enough to melt cadmium. Though people and dogs and lizards and dolphins couldn't survive this little rough patch, maybe some microbes could, by taking to the air (as some folks think may have happened on Venus) or retreating deep underground.

But this middle-aged solar snit is trifling compared to how nasty our star will get in its dotage. In about 5 billion years, the sun will run out of hydrogen fuel and begin morphing into a terrible bloated beast called a red giant.[1] Our formerly happy yellow fellow will swell like a blood-engorged tick, reaching about 250 times its current size—so huge that it engulfs and incinerates Mercury and Venus, and probably Earth as well. (Kids, don't draw a smiley face on a red-giant sun.) Even if our planet manages to avoid complete destruction, it'll get charred like a toddler-toasted campfire marshmallow. This will certainly be the end for life on Earth.

But not necessarily for humanity. We could bound outward like a grasshopper fleeing a wildfire, leaping from Earth to Mars to Europa to Enceladus to Pluto in succession as the sun's Hulk turn warmed these worlds enough to support fragile meatsacks such as us. (I know,

---

1   By the way, Earth's nighttime sky will look very different by this point than it does today. In about 4 billion years, our own Milky Way will merge with the huge neighboring Andromeda spiral, forming a hybrid beast astronomers have dubbed Milkdromeda, presumably because AndroWay sounds too much like a bodybuilding supplement. Our solar system likely won't feel any significant direct effects from this collision; the spaces between stars in both galaxies are so vast that few of them will interact.

we may have sublimated ourselves to virtual or digital meatsacks by then, but putting our fleshy, hairy selves into this scenario is more relatable. So let's go with it.) And then we could keep going, traversing the depths of interstellar space to other star systems, buying billions or even trillions of years of prestige-TV-watching time along the way.

That's right: trillions. Because, as you may recall from chapter 6, that's how long red dwarfs, the most common stars in the Milky Way, keep on shining. (How long stars live is a function of their initial mass. Giant stars much larger than the sun burn through all their fuel in just a few million years, then die in spectacular supernova explosions.)

When red dwarfs finally burn through all of their fuel, they become super-dense corpses called white dwarfs—the same fate that awaits our sun once it's done turning planets into charcoal briquettes. How dense is super dense? Well, our dead sun will pack about 50 percent of its birth mass into a ball as big as Earth—meaning a teaspoon of the stuff would weigh more than 5 tons, if you could somehow transport it here and find a zoo willing to loan out its elephant scale. (Scooping that little bit up would be a challenge, too: white dwarfs' surface gravity is more than 100,000 times stronger than that of Earth, so you'd need a really long spoon to avoid getting squished.)

White dwarfs are cosmic ghosts: glowing dead things. They slowly radiate away tremendous amounts of built-up heat, so we—and any other aliens who share the cosmos with us—may be able to warm our hands by them for another few trillion years. But even these last lights will fade and blink out. Hundreds of trillions of years from now, the universe will go dark.

That may be the end of us, or whatever we've become by then. Or maybe not, for there will still be black holes. We could theoretically set up shop on the outskirts of these monsters, harnessing their huge gravity to power our space blenders and rock tumblers. And we could do this for a very long time—until all of the protons in the universe have decayed into smaller particles, making it impossible for us, or anything else substantial, to exist in corporeal form. This disappointing milestone will definitely be reached by 10^{64} years from now, and it may come as soon as 10^{34} years, according to Princeton University physicist J. Richard Gott, who lays out this far-future timeline in *Welcome to the Universe*, a 2016 book he co-authored with Neil deGrasse Tyson and Michael Strauss.

It will just be black holes and particles for a long, long time. And then, 10^{100} years from now, particles will become the universe's undisputed kings. All the black holes will have vanished by then, their mighty essence leached away in piddling little blips of radiation over the eons until nothing at all was left. Death by 10^{100} cuts. (That's right: black holes aren't entirely black. The radiation they emit is called Hawking radiation, because Stephen Hawking predicted its existence back in 1974.)

And that will be that. The universe will keep expanding forever, churning out endless frontiers of darkness for invisible specks to streak through on their way from nothing to nowhere. The end. (This is the mainstream view, anyway. Some cosmologists think dramatic expansion will tear the universe apart in a Big Rip long before we get to black hole extinction, and others posit that our aging universe could cycle its way into a rebirth before it bites the dust.)

This depressing little rundown may strike you as silly or pointless. Why speculate about how we can eke out a living billions or trillions or 10^{64}-illions of years from now, when we have almost no shot of making it that long? After all, the fossil record shows that mammal species survive for just a few million years on average, and *Homo sapiens* seems to be bumbling along on a distinctly subpar trajectory. We've been around just 200,000 years, and we may already be on our way out.

I get it. Those are all valid points. But why not imagine a brighter and bolder future for ourselves? Maybe, if we actually think it's possible, we can become the Greys of legend, the creatures that homebody aliens wonder about as they stare up at their exotic night skies: "Where are they? Will they ever come here?"

If I could beam a reply to our cosmic kin across the light-years, I would: we're here, little fella, trapped on this rock called Earth—but hopefully we won't be trapped for long.

Bibliography

Achenbach, Joel. *Captured by Aliens: The Search for Life and Truth in a Very Large Universe*. New York: Simon and Schuster, 1999.

Adams, Douglas. *The Hitchhiker's Guide to the Galaxy*. London: Pan Books, 1979.

Brin, G. D. "The Great Silence: The Controversy Concerning Extraterrestrial Intelligent Life." *Quarterly Journal of the Royal Astronomical Society* 24 (1983): 283–309.

Conway Morris, Simon. *The Runes of Evolution: How the Universe Became Self-Aware*. West Conshohocken, PA: Templeton Press, 2015.

Crowe, Michael J. *The Extraterrestrial Life Debate, 1750–1900*. Cambridge: Cambridge University Press, 1986.

Davies, Paul. *The Eerie Silence: Renewing Our Search for Alien Intelligence*. New York: Houghton Mifflin Harcourt, 2010.

Everett, Allen, and Thomas Roman. *Time Travel and Warp Drives: A Scientific Guide to Shortcuts through Time and Space*. Chicago: University of Chicago Press, 2011.

Gould, Stephen Jay. *Wonderful Life: The Burgess Shale and the Nature of History*. New York: W. W. Norton, 1989.

Krauss, Lawrence. *The Physics of Star Trek*. New York: Basic Books, 1995.

Krissansen-Totton, J., et al. "Disequilibrium Biosignatures over Earth History and Implications for Detecting Exoplanet Life." *Science Advances* 4, no. 1 (2018). DOI: 10.1126/sciadv.aao5747.

Losos, Jonathan. *Improbable Destinies: Fate, Chance, and the Future of Evolution*. New York: Riverhead Books, 2017.

McKay, D., et al. "Search for Past Life on Mars: Possible Relic Biogenic Activity in Martian Meteorite ALH 84001." *Science* 273 (1996): 924–930.

Pinker, Steven. *The Better Angels of Our Nature: Why Violence Has Declined*. New York: Viking, 2011.

Russell, D. A., and R. Séguin. "Reconstruction of the Small Cretaceous Theropod *Stenonychosaurus inequalis* and a Hypothetical Dinosauroid." *Syllogeus* 37 (1982): 1–43.

Sagan, C., and E. E. Salpeter. "Particles, Environments, and Possible Ecologies in the Jovian Atmosphere." *Astrophysical Journal Supplement Series* 32 (1976): 737–755.

Sawyer, Kathy. *The Rock from Mars: A Detective Story on Two Planets*. New York: Random House, 2006.

Schulze-Makuch, Dirk, and David Darling. *We Are Not Alone: Why We Have Already Found Extraterrestrial Life*. Oxford: Oneworld, 2010.

Shermer, Michael. *Why People Believe Weird Things: Pseudoscience, Superstition, and Other Confusions of Our Time*. New York: Henry Holt and Company, 1997.

Shostak, Seth. *Confessions of an Alien Hunter*. New York: National Geographic Books, 2009.

Shostak, S., and I. Almar. "The Rio Scale Applied to Fictional SETI Detections." IAA-02-IAA.9.1.06, 2002, http://www.setileague.org/iaaseti/abst2002/rio2002.pdf.

Solomon, Scott. *Future Humans: Inside the Science of Our Continuing Evolution*. New Haven, CT: Yale University Press, 2016.

Taylor, Travis, and Bob Boan. *Alien Invasion: How to Defend Earth*. Wake Forest, NC: Baen, 2011.

Tegmark, Max. "Parallel Universes." Pp. 459–491 in *Science and Ultimate Reality: Quantum Theory, Cosmology and Complexity*. Cambridge: Cambridge University Press, 2004.

Tyson, Neil deGrasse, Michael A. Strauss, and J. Richard Gott. *Welcome to the Universe: An Astrophysical Tour*. Princeton, NJ: Princeton University Press, 2016.

Vardanyan, M., et al. "Applications of Bayesian Model Averaging to the Curvature and Size of the Universe." *Monthly Notices of the Royal Astronomical Society: Letters* 413, no. 1 (2011): L91–L95.

Webb, Stephen. *If the Universe Is Teeming with Aliens, Where Is Everybody? 75 Solutions to the Fermi Paradox and the Problem of Extraterrestrial Life*, 2nd ed. New York: Springer, 2015.

Weintraub, Alan. *Religions and Extraterrestrial Life: How Will We Deal with It?* New York: Springer, 2014.

Wheeler, John A., with Kenneth Ford. *Geons, Black Holes, and Quantum Foam: A Life in Physics*. New York: W. W. Norton and Company, 1998.

Zubrin, Robert, with Richard Wagner. *The Case for Mars*. New York: Free Press, 1996.

Index

Achenbach, Joel, 157
Active SETI (Active Search for Extra-
    Terrestrial Intelligence), 128, 175–78
Advanced Aviation Threat Identification
    Program (AATIP), 161–64
Alcubierre, Miguel, 201, 202, 203, 204, 226
alien abduction stories, 13, 40, 167
alien appearance, 52–55, 58–64
    dinosauroids, 58–62, 60
    Greys, 40–42, 41, 63
alien biosignature searches, 74–89
alien communication, 133–41
alien engineering projects, 74–76
alien gas, 76–84
alien Greys, 40–42, 41, 63
    microbes vs., 143–45
alien invasion, 168–79
alien megastructure theory, 94
alien sex, 66–73
alien-worshipping cults, 150–51
Allan Hills 84001 (ALH84001), 88, 131–32,
    143–44, 149, 155–61
Allen, Paul, 118
Allen Telescope Array (ATA), 20, 118
Almár, Iván, 126–27
amino acids, 49, 85, 86, 107
ammonia, 49, 81
Andromeda spiral, 230n
anglerfish, 70
antimatter rockets, 16, 171, 206–7
apophallation, 71
appearance of aliens. See alien appearance
archaea, 111

Area 51, 165
Arecibo Message, 136–37, 139
Arecibo Observatory, 89, 136
Arrival (film), 63, 140–41
arsenic, 107
artificial intelligence, 64–65, 213
asexual reproduction, 67
asteroid impacts, 14, 24, 49, 58, 61, 127
asteroid mining, 5, 186–87
Aztecs, 169–70, 172

Bains, William, 48, 50–51, 81, 82
"banana peel mechanism" idea, 226–27
bedbugs, 70
Bell, Jocelyn, 96–97
Benner, Steven, 28–29, 30, 87, 122
Bennu, 117
berserkers, 18, 25
Better Angels of Our Nature, The (Pinker),
    179
Bezos, Jeff, 5, 191
BFR (Big Falcon Rocket), 189–91
Big Bang, 42, 43, 44, 46, 201
Bigelow, Robert, 162
Bigelow Aerospace, 162, 185
Big Rip, 233
biochemistry of life, 47–51, 51
biosignature searches, 74–89
black holes, 131, 227, 232–33
Blue Origin, 5, 191, 195, 197
boron, 30
Bostrom, Nick, 17
Boyajian, Tabetha, 94

Boyajian's star, 94, 96
Bradford, John, 208–9
Breakthrough Listen, 90–91, 95
Breakthrough Starshot, 209–10, 211, 212
Brin, David, 13–14, 25–26
Broad, William, 129
"brown gunk," 113
Bryan, Richard, 19
Buddhism, 148
Burger Time (video game), 17
Burgess Shale, 61

CAESAR (Comet Astrobiology Exploration Sample Return), 117–18
Callisto, 109, 110, 113, 117
canals of Mars, 74–75
Captured by Aliens (Achenbach), 157
carbon-based life, 47–50, *51*
carbon dioxide (CO_2), 75, 79, 80, 81, 104, 111, 192–93
Carl Sagan Institute, 4, 78, 121, 131
carotenoids, 217
Carroll, Lewis, 67*n*
Case for Mars, The (Zubrin), 193–94
Casimir effect, 202
Cassini–Huygens, 110–11, 112, 115, 118
chemofossils, 85
Chicago Pile-1 (CP-1), 9
chloromethane, 81, 155
Christianity, 148–50
"chronology protection conjecture," 226, 227
Clarke, Arthur C., 209
Clinton, Bill, 132, 143, 160
"closed timelike curves," 223
Cocconi, Giuseppe, 139–40
cognitive biases, 59
colonization, 183–99
    of Mars, 188–97
    of moon, 183–88
colony ships, 208, 211–12
Columbus, Christopher, 172
comet impacts, 32, 49, 127
comets, as biodiversity vehicles, 34–37, 49, 174
Confessions of an Alien Hunter (Shostak), 129, 134
conjugation, 68–69
Consolmagno, Guy, 150

contact, 2, 64, 133–41, 179
Contact (Sagan), 11
contamination, 102, 154
contingent evolution, 59, 61–62
convergent evolution, 61–62
Conway Morris, Simon, 59–61
Copernican Revolution, 146–47
cosmic inflation, 44–45, 201
cosmic microwave background, 43
"cosmological constant," 202–3
Cretaceous–Paleogene extinction event, 14, 24, 58, 61
Crick, Francis, 36
crocodile paradox, 12
Crowe, Michael, 99
Curiosity (rover), 27–28, 77, 102, 150, 160
cuttlefish, 54

dark energy, 42, 96, 202–3
DARPA (Defense Advanced Research Projects Agency), 205
Darwin, Charles, 37, 146–47, 152
Davies, Paul, 25
Deep Space Industries, 198
Deinococcus radiodurans, 32–33, 101
desert varnish, 108
digital life, 64–65
dinosauroids, 58–62, *60*
Dione, 109
Di Pippo, Simonetta, 126
"directed panspermia," 36
divergent evolution, 216–17, 220
DNA (deoxyribonucleic acid), 28, 85–87, 137
Dragonfly (drone), 117–18
Drake, Frank, 10, 10*n*, 136–37
Drake equation, 10*n*
Druyan, Ann, 129
Dyson, Freeman, 93–94, 205
Dyson spheres, 93–94, 205

Earth
    extinction events, 14, 24, 58, 61
    future of, 229–34
    history of, 22–24
    origins of life on, 31–39, 80–81, 106–8
Eerie Silence, The (Davies), 25
Ehman, Jerry, 1–3, 4
Elvis, Martin, 187

Enceladus, 24, 34, 87, 98, 109, 110–12, 116, 118
Enceladus Life Signatures and Habitability (ELSAH), 118
Epsilon Eridani, 10
Europa, 24–25, 108–10, 112–14, 117
Europa Clipper, 117
European Extremely Large Telescope, 78
European Space Agency (ESA), 5, 43, 116–17, 183–84
evaporation, 104
Everett, Allen, 226
Everett, Hugh, 45
Everett, Mark Oliver, 45
evolution, 23–24, 37, 47, 52, 59–62, 64, 87, 101, 146–47
    Mars settlers and, 216–21
ExoMars (Exobiology on Mars), 76–77, 85, 116
exoplanet revolution, 10–11
expansion of the universe, 42–44, *46*, 233
extinction events, 14, 24, 58, 61
Extraterrestrial Life Debate, 1750-1900 (Crowe), 99
extremophiles, 101, 107, 111

fake news, 151–52
Farnsworth, Philo, 75
fast radio bursts, 97
fear of aliens, 178–79
Fermi, Enrico, 9–10
Fermi's paradox, 10, 12–13, 21, 25
flat Earthers, 151
floaters, 57–58
formation of solar system, 30
Fortnite (video game), 17
fossil record, 14, 23, 59, 233
founder effect, 216–17
From Earth to the Moon (Verne), 214
fusion rockets, 204–6
Future Humans (Solomon), 216–17, 218, 219
future of Earth, 229–34

Gagarin, Yuri, 199
Gale Crater, 27–28
Galileo (spacecraft), 82–83
Ganymede, 109, 110, 113, 117
gasbags, 57–58
Gauss, Carl, 99, 137–38

general relativity, 222–24, 225–26
genetic mixing, 66–69, 221
Geons, Black Holes, and Quantum Foam (Wheeler), 223
GFAJ-1, 107
Giant Magellan Telescope, 78
Gibson, Everett, 158, 161
GJ 273b, 138, 139, 221
Gödel, Kurt, 223
Goldin, Dan, 161
Goodall, Jane, 24
Gott, J. Richard, 232
Gould, Stephen Jay, 61
grandfather paradox, 225–27
gravity, 54–55, 195, 223
Green Bank Observatory, 10, 89
greenhouse effect, 229
Greys, 40–42, *41*, 63
    microbes vs., 143–45

habitable zones, 11, 23, 60–61, 80, 119–20, 123, 131–32
Hale-Bopp Comet, 151
Hallucigenia, 54
"handedness," 85
Harzan, Jan, 163–64
Hawking, Stephen, 169–70, 175, 179, 226, 227, 233
Hawking radiation, 233
Heaven's Gate, 151
hermaphrodites, 70, 72
Herschel, William, 98–99*n*, 98–100
Herschel Space Observatory, 99*n*
Hewish, Antony, 97
hibernation, 208–9
history of Earth, 22–24
homo sapiens, 219–20
Hoover, Richard, 34–36
horizon problem, 203
Hubble Space Telescope, 77, 114
hydrogen, 81, 111, 112, 118, 186

icy ocean worlds, 108–14
If the Universe Is Teeming with Aliens, Where Is Everybody? (Webb), 12
Improbable Destinies (Losos), 61–62
Incas, 168, 169–70, 172
Independence Day (film), 127, 170
infectious diseases, 169, 173, 218

inflation (cosmology), 44–45, 201
InSight (lander), 160–61
International Academy of Astronautics (IAA), 124–25
International Space Station (ISS), 5, 33, 195, 198, 208, 227–28
Interstellar (film), 228
interstellar exploration, 200–214
    propulsion issues, 200–207
    robotics, 207–11
Io, 109, 113
iSpace, 186

James Webb Space Telescope (JWST), 77–80
Jupiter, 23, 57–58, 82–83, 108–9, 112–14, 117
JUpiter ICy moons Explorer (JUICE), 117

Kaltenegger, Lisa, 4, 78, 79–80, 84, 121, 131, 140
Keats, Jonathon, 147, 152
Kelly, Mark, 228
Kelly, Scott, 227–28
Kepler (spacecraft), 11
Kepler-452b, 54–55
KIC 8462852, 94, 96
Kornienko, Mikhail, 227–28
Krauss, Lawrence, 16
Kurzweil, Ray, 65

language
    decoding challenge, 133–34, 139–40
    math as universal, 136–38
Large Synoptic Survey Telescope (LSST), 95
laser SETI, 20–21, 89, 91
Late Heavy Bombardment, 31–32
Lee, Pascal, 192
Lehnhardt, Kris, 195
Levin, Gil, 153, 155
Lewicki, Chris, 187
life
    biochemistry of, 47–51, *51*
    NASA's definition of, 47, 87
life on Earth, 31–39, 80–81, 106–8
life on Mars, 27–34, 38–39, 100–103, 153–55
    ALH 84001, 88, 131–32, 143–44, 149, 155–61
    subterranean, 101–2
life span, 56

Little Green Man 1 (LGM-1), 97
Loeb, Avi, 95, 177, 179
Logsdon, John, 184, 196
Losos, Jonathan, 61–63
Lowell, Percival, 74–75, 100
Lowell Observatory, 74–75
"lunarians," 99

McGaha, James, 164–65, 166
McKay, Chris, 39, 116, 118, 142, 159
McKay, David, 155–61
Made In Space, 198–99
Mariner 4, 75
Mars
    ALH 84001, 88, 131–32, 143–44, 149, 155–61
    life on. *See* life on Mars
    meteorites from, 29–30, 31–33
    methane mystery, 76–77
    settlements. *See* Mars settlers
    Viking lander biological experiments, 153–55
    water on, 27–29, 74–75
Mars Atmosphere and Volatile EvolutioN (MAVEN), 160
Mars Attacks! (film), 31
Mars-first idea, 27–34, 38–39
Marsketball, 217–18
Mars Oasis, 189–90
Mars settlers (settlements), 5, 188–97
    Bezos and Blue Origin, 5, 191
    evolution and founder effect, 216–21
    Musk and SpaceX, 5, 189–91, 196
Mars Society, 193, 220
Mars 2020, 85, 88, 160
Martians, 19, 74–75
math and alien communication, 136–38
Mauritius, 172
media response to first contact, 128–30
melanin, 217
Messier 13 (M13), 136–37, 139, 152
meteorites, 31, 49
    Allan Hills 84001 (ALH84001), 88, 131–32, 143–44, 149, 155–61
methane, 76–77, 79–80, 81, 82, 86, 101, 111, 116, 152
methanogens, 111, 112
METI (Messaging to Extra-Terrestrial Intelligence), 128, 175–78

METI International, 18–19, 138
microbes, 22–23, 52–53
    alien, 31, 52–53, 100–103
    Greys vs., 143–45
    panspermia, 34–38
    reproduction, 68–69
    "second genesis" of, 26, 39, 106–7
microbiomes, 52–53, 218n
microfossils, 158–59
military doctrine, 172–73
Milky Way, 11, 119, 230n
Millis, Marc, 208, 212
Milner, Yuri, 90–91, 118, 209
Mimas, 109
mining space, 183–88, 198
Mono Lake, 107
Moon Express, 5, 186
moon landing conspiracy theories, 151
moon mining, 185–86
moon outpost, 183–85, 188
Mormonism, 148
Morris, Dick, 157
Morrison, Philip, 139–40
multiverse, 44–45
Musk, Elon, 5, 65, 189–90, 196, 197
Mutual UFO Network, 163–64

NASA (National Aeronautics and Space Administration), 116–17
    biosignature searches, 77–78
    budgetary issues, 19, 118–19
    Mars programs, 5, 85, 102–3, 160–61, 188–89
    space policy directive, 183
natural selection, 23–24, 58, 62, 217. See also evolution
negative energy, 204
Neptune, 50, 109
neutrinos, 21
New Horizons, 205–6
New York Times, 129–31, 144, 162
nitrous oxide, 81
nuclear-fusion rockets, 204–6
nucleobases, 28

Obousy, Richard, 210–11
observable universe, 40, 42–43
octopi, 54

Ohio State University Radio Observatory, 1, 2, 3
Olbers, Heinrich, 99
O'Malley, Jack, 84
Oort Cloud, 34–35
Opportunity (rover), 160
optical (laser) SETI, 20–21, 89, 91
orbital dynamics, 39, 190
organic molecules
    on Mars, 153–54, 156, 158
    tar paradox, 29–30
    on Titan, 115
Orgel, Leslie, 36
origins of life. See life
OSIRIS-REx (Origins Spectral Interpretation Resource Identification Security-Regolith Explorer), 117
Otto, Sarah, 66, 69, 71–72
Oumuamua, 94–96
Outer Space Treaty of 1967, 187
oxygen, 79–81, 86

PANO-SETI, 91
panspermia, 34–38
parallel universe, 44–45
Partial Test Ban Treaty, 205
Peters, Ted, 147–48
Petkowski, Janusz, 81
philosophy, and discovery of alien life, 146
Phoenix (spacecraft), 154–55
phosphates, 30
phosphorus, 107
photosynthesis, 35, 56, 56n, 80–81, 83–84, 196
Physics of Star Trek, The (Krauss), 16
Pima people, 168
Pinker, Steven, 179
Planck (spacecraft), 43
Planet (company), 198
Planetary Resources, 187, 198
plate tectonics, 114
platypus, 54
Pluto, 109, 205
polycyclic aromatic hydrocarbons, 156
Pope, Nick, 135, 165–66, 172–73
prime numbers, 137, 138
Project Blue Book, 162, 164, 166
Project Daedalus, 205
Project Longshot, 205

Project Mogul, 165
Project Orion, 205
Prometheus (film), 36, 38
Proxima b, 78, 121, 209–10
Proxima Centauri, 11–12, 15–16, 55
pulsars, 96–97
p-value, 82
Pythagorean theorem, 137–38

radio telescopes, 20–21, 89–90, 91, 184
Raffles pitcher plant, 53
Reagan, Ronald, 170–71, 171*n*
red dwarfs, 11–12, 119–21, 231
red edge, 83
red giants, 230
Red Queen hypothesis, 67*n*
Reid, Harry, 162
religion, and discovery of alien life, 147–50
Religions and Extraterrestrial Life (Weintraub), 149
rhinos, 53
Richards, Bob, 5
Rio scale, *127*, 127–28
RNA (ribonucleic acid), 28–29
robotics, 207–11
Rock from Mars, The (Sawyer), 157
rogue planets, 121–23
Roman, Thomas, 226
Ross 128 b, 78, 119, 121
Roswell UFO incident, 165
Rowlands, Sherry, 157
Rummel, John, 101, 102–3
Runes of Evolution, The (Conway Morris), 60–61
Russell, Dale, 58–59
Ruvkun, Gary, 35, 36–37, 85–86, 196, 214
Ryle, Martin, 97

Sagan, Carl, 11, 57, 82, 104, 129, 136–37, 159
Sagittarius A*, 131
Salpeter, Edwin, 57
Saturn, 50, 109, 110–11
Sawyer, Kathy, 157
Schulze-Makuch, Dirk, 21, 56, 88, 104–5, 158–59
Science (journal), 132, 157
Seager, Sara, 81–82
sea turtles, 73

"second genesis" of microbes, 26, 39, 106–7
SETI (search for extraterrestrial intelligence)
    budgetary issues, 19–20, 90–91, 118–19
    chances of detection, 89–97
    communication, 133–41
    current strategy, 20–21
    Drake and, 10, 10*n*
    principles for disseminating information, 124–32
SETI Institute, 11, 20, 65, 101, 129
SETI Permanent Committee, 126
SETI Research Center, 89, 91
sexual reproduction, 66–73
"shadow biosphere," 106–8
shape of the universe, 42–44
Shermer, Michael, 167
Shostak, Seth, 65, 123
    communication issues, 134
    first contact, 141, 174, 178, 179
    June 1997 signal, 129–30
    Rio scale, 126, 127–28
silicon-based life, 50–51
simulation hypothesis, 17–18
singularity, 65, 213
sinkers, 57–58
Slipher, Vesto, 75, 79
slugs, 70–71
Smith, Will, 170
snakes, 53, 70, 83
Solar and Heliospheric Observatory (SOHO), 129
solar radiation, 98*n*, 99–100
solar sails, 209–10
Solomon, Scott, 216–17, 218, 219
space mining, 183–88, 198
space sailing, 209–10
SpaceShipTwo, 197–98
space-time, 42–43, 200–202, 222–24
space travel. *See* interstellar exploration
SpaceWorks Enterprises, 208
SpaceX, 5, 189–90, 195, 197–98
Spanish colonization of the Americas, 169–70, 172, 173–74
special relativity, 44, 200–201, 224
speed of light, 16, 44, 200, 201, 203, 204
spiders, 70
Star Trek, 16–17, 18, 50, 63, 171, 201, 206
Strauss, Michael, 232

Streep, Meryl, 29
Sun, 229–32
sunspots, 99–100
Sutter, Paul, 226, 227

Tabby's star, 94, 96
tardigrades, 32–33
tar paradox, 29–30
Tarter, Jill, 11, 20, 26
    active SETI, 176
    first contact, 145
    June 1997 signal, 129
    Rio scale, 126–27
    search for technosignatures, 93, 140
Tau Ceti, 10
Tau Zero Foundation, 208
technosignatures, 93
Tegmark, Max, 42, 44–45
terraforming, 192–94, 196
Theia, 30
Thirty Meter Telescope (TMT), 78
Thomas-Keprta, Kathie, 157–58
thorium, 185
tidal heating, 109
tidal locking, 120
time dilation, 224, 227, 228
time travel, 222–28
Time Travel and Warp Drives (Everett and Roman), 226
Tipler, Frank, 223–24
Titan, 50, 56, 109, 115–16, 117–18
Torino scale, 127
torpor, 208–9
transduction, 68–69
transformation, 68–69
TRAPPIST-1, 11–12, 55, 78, 93, 120, 121
traumatic insemination, 70
Triton, 50, 109
Troodon, 58, 62
Trump, Donald, 183
2001: A Space Odyssey (film), 127–28
Tyson, Neil deGrasse, 232

UFOs (unidentified flying objects), 161–67
ultraviolet radiation (UV), 56, 77, 83–84
unidentified aerial phenomena, 162
United Nations (UN), 125–26
United Nations Office for Outer Space Affairs (UNOOSA), 126–27

Uranus, 98
USS *Nimitz* UFO incident, 161–64, 165

Vakoch, Douglas, 18–19, 138, 176, 177–78, 179
Varnum, Michael, 143–44
Vatican Observatory, 150
Venera 13, 103
Venter, Craig, 85–86
Venus, 103–5, 229
Venus clouds, 104–5
Verne, Jules, 213–14
Viking lander biological experiments, 153–55, 160
Viking program, 76, 88, 102, 128
Virgin Galactic, 197–98, 199
viruses, 68, 174
von der Dunk, Frans, 188
Voyager program, 210

warp drives, 201–3
water, 102–3, 192–93
    carbon-water life, 47–50, *51*
    icy ocean worlds, 28–29, 37, 108–14
"water holes," 2
Webb, Stephen, 12
Weintraub, David, 149
Welcome to the Universe (Gott, Strauss and Tyson), 232
Werthimer, Dan
    contact risks, 145, 175, 178, 179
    decoding issues, 134, 136, 140
    signs of intelligence, 89, 90, 91, 92–93, 96
Wheeler, John, 223
"Where is everybody?", 9–10, 12–13
white dwarfs, 231–32
Whitesides, George, 199
Why People Believe Weird Things (Shermer), 167
Wide Field Infrared Survey Telescope, 78
Wilkinson Microwave Anisotropy Probe, 43
Wonderful Life (Gould), 61
World View Enterprises, 197–98
wormholes, 203–4, 224–25
Wow! signal, 1–3, 129–30
writing systems, 133–34

Zubrin, Robert, 193–94, 196, 220
Zurbuchen, Thomas, 84, 146

Acknowledgments

This book would not have been possible without the help and support of many people, and I'm grateful to them all. First mention must go to the scientists, engineers, policy experts, NASA officials, and other folks who took time out of their busy days to talk to me about warp drives and aliens' sex lives. There are too many to list here, but you can find their names peppered throughout the book.

Matt Latimer and Dylan Colligan at Javelin planted the seed of this book and nurtured it the entire way through. The people at Hachette took a chance on me and worked hard to make *Out There* better, particularly by reining in my worst impulses.[1] Thanks to the entire

---

1   You wouldn't believe how many footnotes I put in the first draft.

Hachette team who worked on the book — Gretchen Young, Katherine Stopa, Yasmin Mathew, Linda Duggins, and Joe Benincase.

Karl Tate made *Out There* much better as well, primarily by providing the illustrations that liven up the book. He also flagged sections of the manuscript that needed more work—places where my explanations were too muddled or muddy-headed. I'm very grateful.

Thanks as well to my colleagues at Space.com, especially managing editor Tariq Malik, for being so understanding and accommodating during the writing of this book. Tariq, a lover of bad puns like myself, also introduced me to an alternate pronunciation for the Saturn moon Enceladus that I ask readers to use in their heads: "Enchiladas."

My family has always been incredibly supportive no matter what weirdo pursuit I've indulged, from tracking rattlesnakes through the Arizona desert to becoming a journalist just as the field began hemorrhaging jobs. Mom, Dad, Sarah, Rob, Taylor, Payton, and Jake: Thanks so much for everything. I love you guys.

And Teddy: I've saved the best for last. I love you, Little Bear. You are my heart.

About the Author

Michael Wall is uniquely qualified to write a snakes-in-space story, should the need ever arise. Before becoming a journalist, he worked as a biologist, studying primarily reptiles and amphibians. He holds a Ph.D. in biology from the University of Sydney and a graduate certificate in science writing from the University of California, Santa Cruz. He has written for Space.com since 2010.